T0211969

SpringerBriefs in Applied Sciences and Technology

PoliMI SpringerBriefs

More information about this subseries at https://link.springer.com/bookseries/11159
http://www.polimi.it

Matteo Sangiorgio · Fabio Dercole ·
Giorgio Guariso

Deep Learning in Multi-step Prediction of Chaotic Dynamics

From Deterministic Models to Real-World Systems

POLITECNICO
MILANO 1863

Matteo Sangiorgio (iD)
Dipartimento di Elettronica,
Informazione e Bioingegneria (DEIB)
Politecnico di Milano
Milan, Italy

Fabio Dercole (iD)
Dipartimento di Elettronica,
Informazione e Bioingegneria (DEIB)
Politecnico di Milano
Milan, Italy

Giorgio Guariso (iD)
Dipartimento di Elettronica,
Informazione e Bioingegneria (DEIB)
Politecnico di Milano
Milan, Italy

ISSN 2191-530X   ISSN 2191-5318 (electronic)
SpringerBriefs in Applied Sciences and Technology
ISSN 2282-2577   ISSN 2282-2585 (electronic)
PoliMI SpringerBriefs
ISBN 978-3-030-94481-0   ISBN 978-3-030-94482-7 (eBook)
https://doi.org/10.1007/978-3-030-94482-7

This Springer imprint is published by the registered company Springer Nature Switzerland AG
The registered company address is: Gewerbestrasse 11, 6330 Cham, Switzerland

Non ti sia grave il fermarti alcuna volta a vedere nelle macchie de' muri, o nella cenere del fuoco, o nuvoli o fanghi, od altri simili luoghi, ne' quali, se ben saranno da te considerati, tu troverai invenzioni mirabilissime... perchè nelle cose confuse l'ingegno si desta.

(It would not be too much of an effort to pause sometimes to look into these stains on walls, the ashes from the fire, the clouds, the mud, or other similar places. If these are well contemplated, you will find fantastic inventions... These will do you well because they will awaken genius.)

Leonardo da Vinci, Trattato della pittura, ca. 1540.

Preface

In the present data-rich era, we know that time series of many variables can hardly be interpreted as regular movements plus some stochastic noise. For half a century, we have also known that even apparently simple sets of nonlinear equations can produce extremely complex movements that remain within a limited portion of the variables space without being periodic. Such movements have been named "chaotic" ("deterministic chaos" when the equations include no stochasticity).

Immediately after they were discovered, Lorenz and other researchers were troubled by the problem of predictability. How far into the future can we reliably forecast the output of such systems? For many years, the answer to such a question remained limited to very few steps. Today, however, powerful computer tools are available and have been successfully used to accomplish complex tasks. Can we extend our predictive ability using such tools? How far? Can we predict not just a single value, but also an entire sequence of outputs?

This book tries to answer these questions by using deep artificial neural networks as the forecasting tools and analyzing the performances of different architectures of such networks. In particular, we compare the classical feed-forward (FF) architecture with the more recent long short-term memory (LSTM) structure. For the latter, we explore the possibility of using or not the traditional training approach known as "teacher forcing".

Before presenting these methods in detail, we revise the basic elements and tools of chaos theory and nonlinear time-series analysis in Chap. 2. We take a practical approach, looking at how chaoticity can be quantified by simulating mathematical models, as well as from real-world measurements.

Chapter 3 presents the cases on which we test our deep neural predictors. We consider four well-known dynamical systems (in discrete time) showing a chaotic attractor in their variables space: the logistic and the Hénon maps, which are the prototypes of chaos in non-reversible and reversible systems, respectively, and two generalized Hénon maps, to include cases of low- and high-dimensional hyperchaos. These systems easily produce arbitrarily long synthetic datasets, in the form of deterministic chaotic time series, to which we also add a synthetic noise to better mimic real situations. We finally consider two real-world time series: solar irradiance and ozone concentration, measured at two stations in Northern Italy. These time-series are shown to contain a chaotic movement by means of the tools of nonlinear time-series analysis.

Chapter 4 illustrates the structures of the neural networks that we use and introduces their performance metrics. It deals, in particular, with the issue of multi-step forecasting that is commonly performed by recursively nesting one-step-ahead predictions (recursive predictor). An alternative explored here consists of training the model to directly compute multiple outputs (multi-output predictor), each representing the prediction at a specific time step in the future. Multi-output prediction can be implemented in static networks (using FF nets), as well as in a dynamic way (adopting recurrent architectures, such as the LSTM nets). The standard training procedures for both architectures are revised, with particular attention to the issue related to the "teacher forcing" training approach for LSTM networks.

Chapter 5 illustrates the key results of the study on both the synthetic and the real-world time series and explores the effect of different sources of noise. In particular, we consider a stochastic environment that mimics the observation noise, and the presence of non-stationary dynamics that can be seen as a structural disturbance. Additionally, in the context of chaotic systems' forecasting, we introduce the concept of the generalization capability of the neural predictors in terms of "domain adaptation".

Chapter 6 discusses some remarks on the choice of the experimental settings that are adopted in this work. It also presents additional aspects of the neural predictors, analyzing their training method, their neural architecture, and their long-term performances.

The main result we can draw from our journey in the vast area of deep neural network architectures is that the LSTM networks—which are dynamical systems themselves—trained without teacher forcing, are the best approach to predict complex oscillatory time series. They systematically outperform the competitors and also prove to be able to adapt to other domains with similar features without a relevant decrease of accuracy. Overall, they represent a significant improvement of forecasting capabilities of chaotic time series and can be used as components of advanced control techniques, such as model predictive control, to optimize the management of complex real-world systems. These conclusions and some considerations about the expected future research in the field are summarized in the last chapter.

The book will be useful for researchers and Ph.D. students in the area of neural networks and chaotic systems as well as to practitioners that are interested in applying modern deep learning techniques to the forecasting of complex real-world time series, particularly those related to environmental systems. The book will also be of interest for all the scholars that would like to bridge the gap between the classical theory of nonlinear systems and recent development in machine learning techniques.

The book largely draws from the Ph.D. thesis in Information Technology by Matteo Sangiorgio; supervisor: Giorgio Guariso; tutor: Fabio Dercole, Politecnico di Milano, 2021. The main results presented in the text appeared in the following publications:

- M. Sangiorgio and F. Dercole. "Robustness of LSTM neural networks for multi-step forecasting of chaotic time series". *Chaos, Solitons & Fractals* 139, 2020.
- G. Guariso, G. Nunnari and M. Sangiorgio. "Multi-step solar irradiance forecasting and domain adaptation of deep neural networks". In: *Energies* 13.15, 2020.
- F. Dercole, M. Sangiorgio and Y. Schmirander. "An empirical assessment of the universality of ANNs to predict oscillatory time series". *IFAC-PapersOnLine* 53, proceedings of IFAC World Congress, Berlin, Germany, July 2020.
- M. Sangiorgio, F. Dercole and G. Guariso. "Sensitivity of chaotic dynamics prediction to observation noise". *IFAC-PapersOnLine* 54, proceedings of IFAC CHAOS Conference, Catania, Italy, September 2021.
- M. Sangiorgio, F. Dercole and G. Guariso. "Forecasting of noisy chaotic systems with deep neural networks". *Chaos, Solitions and Fractals* 153, 2021.

Milan, Italy
September 2021

Matteo Sangiorgio
Fabio Dercole
Giorgio Guariso

Contents

1   **Introduction to Chaotic Dynamics' Forecasting** 1
    References 5

2   **Basic Concepts of Chaos Theory and Nonlinear Time-Series**
    Analysis 11
    2.1   Dynamical Systems and Their Attractors 12
    2.2   Lyapunov Exponents 14
        2.2.1   Average Exponents 16
        2.2.2   Local Exponents 18
    2.3   Chaotic Systems, Predictability, and Fractal Geometry 19
        2.3.1   Chaotic Attractors and Lyapunov Time Scale 19
        2.3.2   Correlation and Lyapunov Dimensions 21
    2.4   Attractor Reconstruction from Data 23
        2.4.1   Delay-Coordinate Embedding 24
        2.4.2   Estimation of the Largest Lyapunov Exponent 27
    References 28

3   **Artificial and Real-World Chaotic Oscillators** 31
    3.1   Artificial Chaotic Systems 31
        3.1.1   Logistic Map 32
        3.1.2   Hénon Map 33
        3.1.3   Generalized Hénon Map 33
        3.1.4   Time-Varying Logistic Map 34
    3.2   Real-World Time Series 35
        3.2.1   Solar Irradiance 35
        3.2.2   Ozone Concentration 37
    References 40

4   **Neural Approaches for Time Series Forecasting** 43
    4.1   Neural Approaches for Time Series Prediction 44
        4.1.1   FF-Recursive Predictor 45
        4.1.2   FF-Multi-Output Predictor 46

       4.1.3 LSTM Predictor 46
  4.2 Performance Metrics 52
  4.3 Training Procedure 55
  References 56

5 Neural Predictors' Accuracy 59
  5.1 Deterministic Systems 59
       5.1.1 Performance Distribution over the System's Attractor 64
       5.1.2 Sensitivity to the Embedding Dimension 65
  5.2 Stochastic Time Series 67
  5.3 Non-Stationary System 71
  5.4 Real-World Study Cases 73
       5.4.1 Solar Irradiance 73
       5.4.2 Ozone Concentration 78
  References 83

6 Neural Predictors' Sensitivity and Robustness 85
  6.1 Simplicity and Robustness of the Experimental Setting 85
  6.2 Predictors' Long-Term Behavior 87
  6.3 Remarks on the Training Procedure 90
       6.3.1 Backpropagation and Backpropagation Through Time 90
       6.3.2 Training with and Without Teacher Forcing 93
  6.4 Advanced Feed-Forward Architectures 94
  6.5 Chaotic Dynamics in Recurrent Networks 95
  References 95

7 Concluding Remarks on Chaotic Dynamics' Forecasting 97
  References 101

Index 103

Chapter 1
Introduction to Chaotic Dynamics' Forecasting

Abstract Chaotic dynamics are the paradigm of complex and unpredictable evolution due to their built-in feature of amplifying arbitrarily small perturbations. The
forecasting of these dynamics has attracted the attention of many scientists since
the discovery of chaos by Lorenz in the 1960s. In the last decades, machine learning techniques have shown a greater predictive accuracy than traditional tools from
nonlinear time-series analysis. In particular, artificial neural networks have become
the state of the art in chaotic time series forecasting. However, how to select their
structure and the training algorithm is still an open issue in the scientific commu-
nity, especially when considering a multi-step forecasting horizon. We implement
feed-forward and recurrent architectures, considering different training methods and
forecasting strategies. The predictors are evaluated on a wide range of problems,
from low-dimensional deterministic cases to real-world time series.

Since Edward Lorenz's discovery of deterministic chaos in the sixties [63], many
attempts to forecast the future evolution of chaotic systems and to discover how
far they can be predicted, have been carried out adopting a wide range of models.
Some early attempts were performed in the 80s and 90s [7, 18, 31, 50, 75, 82–
84, 91, 97], but the topic has been debated more and more in recent years (see
Fig. 1.1) due to the development of lots of machine learning techniques in the field
of time-series analysis and prediction. Attempts include widely used models such
as the support vector machines [41, 73, 114] and the polynomial functions [92,
93], but the most widely used architectures are artificial neural networks (ANNs).
The main distinction within ANNs is between feed-forward (FF) structures—static
maps which approximate the relationship between inputs and outputs—and recurrent
neural nets (RNNs)—dynamical models whose outputs also depend on an internal
state characterizing each neuron.

Examples of the first category are the traditional multi-layer perceptrons [4, 9,
24, 25, 28, 42, 52, 56, 61, 103, 110], radial basis function networks [26, 38, 57, 71,
95, 96, 104], fuzzy neural networks [22, 33, 53, 69, 108, 117], deep belief neural
nets [54], extreme learning machines [30, 72], and convolutional neural networks
[8, 81]. RNNs are particularly suited to be used as predictors of nonlinear time series

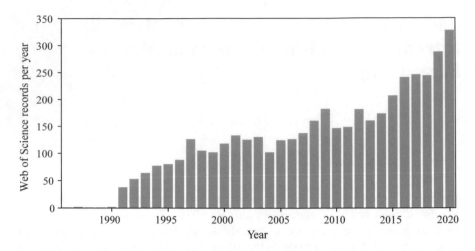

Fig. 1.1 Number of records by publication year (1986–2020) from a Web of Science search for the topic "chaos prediction"

[15, 16, 19, 20, 43, 61, 66, 67, 113, 116], though their training is computationally heavy because of the back propagation through time of the internal memory [36]. Recently, two particular RNN classes, reservoir computers and long short-term memory (LSTM) networks, have been demonstrated to be extremely efficient in the prediction of chaotic dynamics [3, 10, 13, 14, 29, 40, 44, 49, 51, 64, 65, 68, 74, 79, 80, 87, 98–102, 111, 115, 118]. Reservoir computers make use of many recurrent neurons with random connections and training is only performed on the output layer. Conversely, recurrent neurons are trained in LSTM nets. These nets avoid the typical issue of vanishing or exploding optimization gradients by modulating the information flow through so-called gates, whose behaviors are learned as well.

The comparison between feed-forward and recurrent nets is not trivial. Despite the initial optimism, RNNs turned out to achieve similar performances to traditional FF architectures on many forecasting tasks [34, 67]. Recurrent neurons have been demonstrated to be efficient when they are used as basic blocks to build up sequence-to-sequence architectures. This kind of structure represents the state-of-the-art approach in many natural language processing tasks (e.g., machine translation, question answering, automatic summarization, text-to-speech, ...), though they have recently lost ground to advanced FF architectures. Other works in the field of chaotic dynamics' forecasting focused on wavelet networks [17], swarm-optimized neural nets [62], Chebyshev [2], and residual [67] nets. Some authors developed hybrid forecasting schemes that integrate machine learning tools and knowledge-based models [27, 55, 58, 78, 98] or combine multiple predictive techniques [23, 47, 76, 90].

Forecasting chaotic dynamics one or few time steps ahead is usually an easy task as demonstrated by the high performances obtained in many systems, in both continuous and discrete time [1, 4, 5, 59, 68, 69, 103, 107, 109, 116]. The situation changes dramatically when considering a mid-long horizon, because of the expansion of small errors intrinsically due to the chaotic nature of the actual data. Even when the data are generated by a known chaotic system, small initialization or numerical errors are amplified up to predict a distant point in the system's attractor.

The forecasting of a time series over a multi-step horizon is unequivocally recognized as an open issue in the field of time series forecasting and a challenging testing ground for machine learning techniques [46, 70].

The multi-step ahead forecasting of a chaotic dynamic is commonly done by recursively performing one-step ahead predictions [2, 6, 32, 60, 89, 105]. An alternative consists of training the model to directly compute multiple outputs [96], each of which represents the prediction at a certain time step, or even identifying a specific model for each future step [62].

In this book, we compare three state-of-the-art architectures for the multi-step prediction of chaotic time series: FF-recursive, FF-multi-output, and LSTM nets, the latter trained according to two different methods: the so-called teacher forcing (TF), so far traditionally used, and the variant without teacher forcing (no-TF). We analyze the strengths and weaknesses of these four predictors, from both a qualitative and quantitative perspective.

Because they are optimized to predict only the next time step, the FF-recursive net is trained to replicate the dynamical systems generating the data, i.e., the map from previous data to the next one. Its performance, however, degrades on mid-long horizons because of the system's chaoticity. The FF-multi-output net is trained to predict the whole forecasting horizon, but it acts as if the outputs were different variables, rather than the same variable sampled at subsequent steps. For this reason, it struggles to replicate the dynamical behavior of the chaotic maps.

To overcome the critical aspects of FF nets, we adopt models which are able to deal with temporal dynamics, such as the LSTM nets. These architectures can potentially couple the advantages of the FF-recursive and multi-output nets. However, they are usually trained with the so-called TF algorithm, in which the target values are used as an input for each time step ahead, rather than the outputs predicted for the previous steps. Essentially, the net learns to predict only one step, because even when predicting further, it receives the actual data up to the step before the one to be predicted. In other words, the TF prevents small errors in the approximated map from being corrected, since they are not propagated in time. In inference mode, multi-step predictions necessarily require the previous prediction to be fed back, so that the network behaviors in training and prediction do not coincide (this issue is known as exposure bias in the machine learning literature [45, 85]). TF is critical, especially when dealing with chaotic dynamics for their well-known sensitivity to initialization and prediction errors.

To solve this issue, we propose abandoning the TF method, feeding back the network predictions even during training. We therefore compare four neural architectures: FF-recursive, FF-multi-output, LSTM-TF, and LSTM-no-TF.

The articles in the field of chaotic dynamics' forecasting, except for a few recent exceptions [12, 21, 77, 88, 94], usually study noise-free stationary time series. We analyze how the predictors' forecasting skills are affected by different typologies of noise. To quantify the effect of the observation noise, we artificially add a stochastic disturbance with different magnitudes to the deterministic time series generated by the traditional chaotic systems. We also assess the sensitivity to the structural noise, obtained by introducing a slow-varying dynamic for the parameter defining the growth rate in the traditional logistic map. The resulting non-stationary process has concurrent slow and fast dynamics that represent a challenging forecasting task. These numerical experiments lay somewhat in between the deterministic systems, mainly theoretical, and the practical applications. Finally, we apply the proposed methodologies to two real-world time series that exhibit a chaotic behavior: solar irradiance and ozone concentration.

The results obtained show that, in general, LSTM nets trained without TF are the highest performing in predicting chaotic dynamics. The improvement with respect to the three competitors is not uniform and strongly task-dependent. LSTM-no-TF are also shown to be more robust when a redundant (or lacking) number of time lags are included in the input—a feature that represents a remarkable advantage given the well-known problem of estimating the actual embedding dimension from a time series [11].

Another remarkable feature of the forecasting models is their generalization capability, often mentioned as "domain adaptation" (a sub-field of the so-called "transfer learning") in the neural networks literature [35]. It indicates the possibility of storing the knowledge gained while solving a given task and applying it to different, though similar, datasets [112]. Transfer learning became a hot topic in machine learning in the last decade, but it only received attention from researchers in the field of nonlinear science and chaotic systems in recent times [37, 39, 48, 86, 106]. To test this feature, the neural networks developed to forecast the solar irradiance in a specific location (source domain) have been used, without retraining, on other sites (target domains) with quite different geographical conditions. The neural networks developed in our study have proved able to forecast solar radiation in other stations with a minimal loss of precision.

The rest of the book is structured as follows. Chapter 2 introduces some basic concepts related to chaos theory. Chapter 3 presents the chaotic systems and time series used to test the predictors. Chapter 4 describes different ways to use feedforward and recurrent neural networks in a multi-step ahead prediction task. The other sections of this chapter focus on some technical issues on how to set up a supervised learning task starting from a time series and on the performance metrics and the training procedure. Chapter 5 reports the forecasting performance obtained in both the artificial systems and the real-world time series considered. In Chap. 6, we discuss some technical aspects of the training procedure and of the behavior of the different neural predictors. We also present an overview of other alternative state-of-the-art architectures which can be adopted for numerical time series forecasting. Chapter 7 provides the concluding remarks of the book and presents some possible future research directions.

References

1. Abdulkadir, S. J., Alhussian, H., & Alzahrani, A. I. (2018). Analysis of recurrent neural networks for henon simulated time-series forecasting. *Journal of Telecommunication, Electronic and Computer Engineering (JTEC), 10.1-8*, 155–159.
2. Akritas, P., Antoniou, I., & Ivanov, V. V. (2000). Identification and prediction of discrete chaotic maps applying a Chebyshev neural network. *Chaos, Solitons and Fractals, 11.1-3*, 337–344.
3. Antonik, P., et al. (2018). Using a reservoir computer to learn chaotic attractors, with applications to chaos synchronization and cryptography. *Physical Review E, 98.1*, 012215.
4. Atsalakis, G., Skiadas, C., & Nezis, D. (2008). Forecasting Chaotic time series by a Neural Network. In *Proceedings of the 8th International Conference on Applied Stochastic Models and Data Analysis, Vilnius, Lithuania.* (Vol. 30, p. 7782).
5. Atsalakis, G., & Tsakalaki, K. (2012). Simulating annealing and neural networks for chaotic time series forecasting. *Chaotic Model. Simul., 1*, 81–90.
6. Bakker, R., et al. (2000). Learning chaotic attractors by neural networks. *Neural Computation, 12.10*, 2355–2383.
7. Bollt, E. M. (2000). Model selection, confidence and scaling in predicting chaotic time-series. *International Journal of Bifurcation and Chaos, 10.06*, 1407–1422.
8. Bompas, S., Georgeot, B., & Guéry-Odelin, D. (2020). Accuracy of neural networks for the simulation of chaotic dynamics: Precision of training data versus precision of the algorithm. arXiv:2008.04222.
9. Bonnet, D., Labouisse, V., & Grumbach, A. (1997). δ-NARMA neural networks: A new approach to signal prediction. *IEEE Transactions on Signal Processing, 45.11*, 2799–2810.
10. Borra, F., Vulpiani, A., & Cencini, M. (2020). Effective models and predictability of chaotic multiscale systems via machine learning. *Physical Review E, 102.5*, 052203.
11. Bradley, E., & Kantz, H. (2015). Nonlinear time-series analysis revisited. *Chaos: An Interdisciplinary Journal of Nonlinear Science, 25.9*, 097610.
12. Brajard, J. et al. (2020). Combining data assimilation and machine learning to emulate a dynamical model from sparse and noisy observations: A case study with the Lorenz 96 model. *Journal of Computational Science, 44*, 101171.
13. Butcher, J. B., et al. (2013). Reservoir computing and extreme learning machines for nonlinear time-series data analysis. *Neural Networks, 38*, 76–89.
14. Canaday, D., Griffith, A., & Gauthier, D. J. (2018). Rapid time series prediction with a hardware-based reservoir computer. *Chaos: An Interdisciplinary Journal of Nonlinear Science, 28.12*, 123119.
15. Cannas, B., & Cincotti, S. (2002). Neural reconstruction of Lorenz attractors by an observable. *Chaos, Solitons and Fractals, 14.1*, 81–86.
16. Cannas, B., et al. (2001). Learning of Chua's circuit attractors by locally recurrent neural networks. *Chaos, Solitons and Fractals, 12.11*, 2109–2115.
17. Cao, L., et al. (1995). Predicting chaotic time series with wavelet networks. *Physica D: Nonlinear Phenomena, 85.1-2*, 225–238.
18. Casdagli, M. (1989). Nonlinear prediction of chaotic time series. *Physica D: Nonlinear Phenomena, 35.3*, 335–356.
19. Cechin, A. L., Pechmann, D. R., & de Oliveira, L. P. (2008). Optimizing Markovian modeling of chaotic systems with recurrent neural networks. *Chaos, Solitons and Fractals, 37.5*, pp. 1317–1327.
20. Chandra, R., & Zhang, M. (2012). Cooperative coevolution of Elman recurrent neural networks for chaotic time series prediction. *Neurocomputing, 86*, 116–123.
21. Chen, P., et al. (2020). Autoreservoir computing for multistep ahead prediction based on the spatiotemporal information transformation. *Nature Communications, 11.1*, 1–15.
22. Chen, Z. (2010). A chaotic time series prediction method based on fuzzy neural network and its application. In *International Workshop on Chaos-Fractal Theories and Applications* (pp. 355–359). IEEE.

23. Cheng, W., et al. (2021). High-efficiency chaotic time series prediction based on time convolution neural network. *Chaos, Solitons and Fractals, 152*, 111304.
24. Covas, E., & Benetos, E. (2019). Optimal neural network feature selection for spatial-temporal forecasting. *Chaos: An Interdisciplinary Journal of Nonlinear Science, 29.6*, 063111.
25. Dercole, F., Sangiorgio, M., & Schmirander, Y. (2020). An empirical assessment of the universality of ANNs to predict oscillatory time series. *IFAC-PapersOnLine, 53.2*, 1255–1260.
26. Ding, H.-L., et al. (2009). Prediction of chaotic time series using L-GEM based RBFNN. In *2009 International Conference on Machine Learning and Cybernetics* (Vol. 2, pp. 1172–1177). IEEE.
27. Doan, N. A. K., Polifke, W., & Magri, L. (2019). Physics-informed echo state networks for chaotic systems forecasting. In *International Conference on Computational Science* (pp. 192–198). Springer.
28. Dudul, S. V. (2005). Prediction of a Lorenz chaotic attractor using two-layer perceptron neural network. *Applied Soft Computing, 5.4*, pp. 333–355.
29. Fan, H., et al. (2020). Long-term prediction of chaotic systems with machine learning. *Physical Review Research, 2.1*, 012080.
30. Faqih, A., Kamanditya, B., & Kusumoputro, B. (2018). Multi-step ahead prediction of Lorenz's Chaotic system using SOM ELM- RBFNN. In *2018 International Conference on Computer, Information and Telecommunication Systems (CITS)* (pp. 1–5). IEEE.
31. Farmer, J. D., & Sidorowich, J. J. (1987). Predicting chaotic time series. *Physical Review Letters, 59.8*, 845.
32. M Galván, I., & Isasi, P. (2001). Multi-step learning rule for recurrent neural models: An application to time series forecasting. *Neural Processing Letters, 13.2*, 115–133.
33. Gao, Y., & Joo Er, M. (2005). NARMAX time series model prediction: Feedforward and recurrent fuzzy neural network approaches. *Fuzzy Sets and Systems, 150.2*, 331–350.
34. Gers, F. A., Eck, D., & Schmidhuber, J. (2002). Applying LSTM to time series predictable through time-window approaches. In *Neural Nets WIRN Vietri-01* (pp. 193–200). Springer.
35. Glorot, X., Bordes, A., & Bengio, Y. (2011). Domain adaptation for large-scale sentiment classification: A deep learning approach. *Proceedings of the 28 th International Conference on Machine Learning*, Bellevue, WA, USA.
36. Goodfellow, I., Bengio, Y., & Courville, A. (2016). *Deep Learning*. MIT Press.
37. Guariso, G., Nunnari, G., & Sangiorgio, M. (2020). Multi-step solar irradiance forecasting and domain adaptation of deep neural networks. *Energies, 13.15*, 3987.
38. Guerra, F. A., & dos Coelho, L. S. (2008). Multi-step ahead nonlinear identification of Lorenz's chaotic system using radial basis neural network with learning by clustering and particle swarm optimization. *Chaos, Solitons and Fractals, 35.5*, 967–979.
39. Guo, Y., et al. (2020). Transfer learning of chaotic systems. arXiv:2011.09970.
40. Haluszczynski, A., & Räth, C. (2019). Good and bad predictions: Assessing and improving the replication of chaotic attractors by means of reservoir computing. *Chaos: An Interdisciplinary Journal of Nonlinear Science, 29.10*, 103143.
41. Han, L., Ding, L., & Qi, L. (2005). Chaotic Time series nonlinear prediction based on support vector machines. *Systems Engineering - Theory and Practice, 9*.
42. Han, M., & Wang, Y. (2009). Analysis and modeling of multivariate chaotic time series based on neural network. *Expert Systems with Applications, 36.2*, 1280–1290.
43. Han, M. et al. (2004). Prediction of chaotic time series based on the recurrent predictor neural network. *IEEE Transactions on Signal Processing, 52.12*, 3409–3416.
44. Hassanzadeh, P., et al. (2019). Data-driven prediction of a multi-scale Lorenz 96 chaotic system using a hierarchy of deep learning methods: Reservoir computing, ANN, and RNN-LSTM. In *Bulletin of the American Physical Society*, C17-009.
45. He, T., et al. (2019). Quantifying exposure bias for neural language generation. arXiv:1905.10617
46. Hussein, S., Chandra, R., & Sharma, A. (2016). Multi-step- ahead chaotic time series prediction using coevolutionary recurrent neural networks. In *IEEE Congress on Evolutionary Computation (CEC)* (pp. 3084–3091). IEEE.

47. Inoue, H., Fukunaga, Y., & Narihisa, H. (2001). Efficient hybrid neural network for chaotic time series prediction. In *International Conference on Artificial Neural Networks* (pp. 712–718). Springer.
48. Inubushi, M., & Goto, S. (2020). Transfer learning for nonlinear dynamics and its application to fluid turbulence. *Physical Review E, 102.4*, 043301.
49. Jiang, J., & Lai, Y.-C. (2019). Model-free prediction of spatiotemporal dynamical systems with recurrent neural networks: Role of network spectral radius. *Physical Review Research, 1.3*, 033056.
50. Jones, R. D., et al. (1990). Function approximation and time series prediction with neural networks. In *1990 IJCNN International Joint Conference on Neural Networks* (pp. 649–665). IEEE.
51. Jüngling, T. (2019). Reconstruction of complex dynamical systems from time series using reservoir computing. In *IEEE International Symposium on Circuits and Systems (ISCAS)* (pp. 1–5). IEEE.
52. Karunasinghe, D. S. K., & Liong, S.-Y. (2006). Chaotic time series prediction with a global model: Artificial neural network. *Journal of Hydrology, 323.1-4*, 92–105.
53. Kuremoto, T., et al. (2003). Predicting chaotic time series by reinforcement learning. In *Proceedings of the 2nd International Conferences on Computational Intelligence, Robotics and Autonomous Systems* (CIRAS 2003).
54. Kuremoto, T. (2014). Forecast chaotic time series data by DBNs. In *7th International Congress on Image and Signal Processing* (pp. 1130–1135). IEEE.
55. Lei, Y., Hu, J., & Ding, J. (2020). A hybrid model based on deep LSTM for predicting high-dimensional chaotic systems. arXiv:2002.00799.
56. Lellep, M., et al. (2020). Using machine learning to predict extreme events in the Hénon map. In *Chaos: An Interdisciplinary Journal of Nonlinear Science, 30.1*, 013113.
57. Leung, H., Lo, T., & Wang, S. (2001). Prediction of noisy chaotic time series using an optimal radial basis function neural network. *IEEE Transactions on Neural Networks, 12.5*, 1163–1172.
58. Levine, M. E., & Stuart, A. M. (2021). A framework for machine learning of model error in dynamical systems. arXiv:2107.06658.
59. Li, Q., & Lin, R.-C. (2016). A new approach for chaotic time series prediction using recurrent neural network. *Mathematical Problems in Engineering*, 3542898.
60. Lim, T. P., & Puthusserypady, S. (2006). Error criteria for cross validation in the context of chaotic time series prediction. *Chaos: An Interdisciplinary Journal of Nonlinear Science, 16.1*, 013106.
61. Lin, T.-N., et al. (1997). A delay damage model selection algorithm for NARX neural networks. *IEEE Transactions on Signal Processing, 45.11*, 2719–2730.
62. López-Caraballo, C. H., et al. (2016). Mackey-Glass noisy chaotic time series prediction by a swarm-optimized neural network. *Journal of Physics: Conference Series, 720*, 1. IOP Publishing.
63. Lorenz, E. N. (1963). Deterministic nonperiodic flow. *Journal of the Atmospheric Sciences, 20.2*, 130–141.
64. Lu, Z., Hunt, B. R., & Ott, E. (2018). Attractor reconstruction by machine learning. *Chaos: An Interdisciplinary Journal of Nonlinear Science, 28.6*, 061104.
65. Lu, Z., et al. (2017). Reservoir observers: Model-free inference of unmeasured variables in chaotic systems. *Chaos: An Interdisciplinary Journal of Nonlinear Science, 27.4*, 041102.
66. Ma, Q.-L. (2007). Chaotic time series prediction based on evolving recurrent neural networks. In *International Conference on Machine Learning and Cybernetics* (Vol. 6, pp. 3496–3500). IEEE.
67. Maathuis, H. et al. (2017). Predicting chaotic time series using machine learning techniques. In *Preproceedings of the 29th Benelux Conference on Artificial Intelligence (BNAIC 2017)* (pp. 326–340).
68. Madondo, M., & Gibbons, T. (2018). Learning and modeling chaos using lstm recurrent neural networks. *MICS 2018 Proceedings Paper 26*.

69. Maguire, L. P., et al. (1998). Predicting a chaotic time series using a fuzzy neural network. *Information Sciences, 112.1-4*, 125–136.
70. Mariet, Z., & Kuznetsov, V. (2019). Foundations of sequence-to-sequence modeling for time series. In *The 22nd International Conference on Artificial Intelligence and Statistics*, 408–417.
71. Masnadi-Shirazi, M., & Subramaniam, S. (2020). Attractor Ranked Radial Basis function network: A nonparametric forecasting Approach for chaotic Dynamic Systems. *Scientific Reports, 10.1*, 1–10.
72. Xin-Ying, W. Min, H. (2012). Multivariate chaotic time series prediction based on extreme learning machine. *Acta Physica Sinica, 8*.
73. Mukherjee, S., Osuna, E., & Girosi, F. (1997). Nonlinear prediction of chaotic time series using support vector machines. In *Neural Networks for Signal Processing VII. Proceedings of the 1997 IEEE Signal Processing Society Workshop* (pp. 511–520). IEEE.
74. Nakai, K., & Saiki, Y. (2019). Machine-learning construction of a model for a macroscopic fluid variable using the delay-coordinate of a scalar observable. arXiv:1903.05770.
75. Navone, H. D., & Ceccatto, H. A. (1995). Learning chaotic dynamics by neural networks. *Chaos, Solitons and Fractals, 6*, 383–387.
76. Okuno, S., Aihara, K., & Hirata, Y. (2019). Combining multiple forecasts for multivariate time series via state-dependent weighting. *Chaos: An Interdisciplinary Journal of Nonlinear Science, 29.3*, 033128.
77. Patel, D., et al. (2021). Using machine learning to predict statistical properties of non-stationary dynamical processes: System climate, regime transitions, and the effect of stochasticity. *Chaos: An Interdisciplinary Journal of Nonlinear Science, 31.3*, 033149.
78. Pathak, J., et al. (2018). Hybrid forecasting of chaotic processes: Using machine learning in conjunction with a knowledge-based model. *Chaos: An Interdisciplinary Journal of Nonlinear Science, 28.4*, 041101.
79. Pathak, J., et al. (2018). Model-free prediction of large spatiotemporally chaotic systems from data: A reservoir computing approach. *Physical Review Letters, 120.2*, 024102.
80. Pathak, J., et al. (2017). Using machine learning to replicate chaotic attractors and calculate Lyapunov exponents from data. *Chaos: An Interdisciplinary Journal of Nonlinear Science, 27.12*, 121102.
81. Penkovsky, B., et al. (2019). Coupled nonlinear delay systems as deep convolutional neural networks. *Physical Review Letters, 123.5*, 054101.
82. Principe, J. C., & Kuo, J.-M. (1995). Dynamic modelling of chaotic time series with neural networks. *Proceedings of the 7th International Conference on Neural Information Processing Systems*, 311–318.
83. Principe, J. C., Rathie, A., Kuo, J.-M. (1992). Prediction of chaotic time series with neural networks and the issue of dynamic modeling. *International Journal of Bifurcation and Chaos, 2.04*, 989–996.
84. Principe, J. C., Wang, L., & Kuo, J.-M. (1998). Non-linear dynamic modelling with neural networks. In *Signal Analysis and Prediction* (pp. 275–290). Springer.
85. Ranzato, M., et al. (2015). Sequence level training with recurrent neural networks. arXiv:1511.06732.
86. Sangiorgio, M. (2021). Deep learning in multi-step forecasting of chaotic dynamics. Ph.D. thesis. Department of Electronics, Information and Bioengineering, Politecnico di Milano.
87. Sangiorgio, M., & Dercole, F. (2020). Robustness of LSTM neural networks for multi-step forecasting of chaotic time series. *Chaos, Solitons and Fractals, 139*, 110045.
88. Sangiorgio, M., Dercole, F., & Guariso, G. (2021). Forecasting of noisy chaotic systems with deep neural networks. *Chaos, Solitons & Fractals, 153*, 111570.
89. Shi, X., et al. (2017). Chaos time-series prediction based on an improved recursive Levenberg-Marquardt algorithm. *Chaos, Solitons and Fractals, 100*, 57–61.
90. Shi, Z., & Han, M. (2007). Support vector echo-state machine for chaotic time-series prediction. *IEEE Transactions on Neural Networks, 18.2*, 359–372.
91. Shukla, J. (1998). Predictability in the midst of chaos: A scientific basis for climate forecasting. *Science, 282.5389*, 728–731.

92. Su, L., & Li, C. (2015). Local prediction of chaotic time series based on polynomial coefficient autoregressive model. *Mathematical Problems in Engineering*, 901807.
93. Su, L.-Y. (2010). Prediction of multivariate chaotic time series with local polynomial fitting. *Computers and Mathematics with Applications, 59.2*, 737–744.
94. Teng, Q., & Zhang, L. (2019). Data driven nonlinear dynamical systems identification using multi-step CLDNN. *AIP Advances, 9.8*, p. 085311.
95. Todorov, Y., Koprinkova-Hristova, P., & Terziyska, M. (2017). Intuitionistic fuzzy radial basis functions network for modeling of nonlinear dynamics. In *2017 21st International Conference on Process Control (PC)* (pp. 410–415). IEEE.
96. Van Truc, N., & Anh, D. T. (2018). Chaotic time series prediction using radial basis function networks. *2018 4th International Conference on Green Technology and Sustainable Development (GTSD)* (pp. 753–758). IEEE.
97. Verdes, P. F., et al. (1998). Forecasting chaotic time series: Global versus local methods. *Novel Intelligent Automation and Control Systems, 1*, 129–145.
98. Vlachas, P. R., et al. (2018). Data-driven forecasting of high-dimensional chaotic systems with long short-term memory networks. In *Proceedings of the Royal Society A: Mathematical, Physical and Engineering Sciences* (Vol. 474.2213, p. 20170844).
99. Vlachas, P. R. et al. (2020). Backpropagation algorithms and Reservoir Computing in Recurrent Neural Networks for the forecasting of complex spatiotemporal dynamics. *Neural Networks, 126*, 191–217.
100. Wan, Z. Y., et al. (2018). Data-assisted reduced-order modeling of extreme events in complex dynamical systems. *PLoS One, 13.5*, e0197704.
101. Wang, R., Kalnay, E., & Balachandran, B. (2019). Neural machine-based forecasting of chaotic dynamics. *Nonlinear Dynamics, 98.4*, 2903–2917.
102. Weng, T., et al. (2019). Synchronization of chaotic systems and their machine-learning models. *Physical Review E, 99.4*, 042203.
103. Woolley, J. W., Agarwal, P. K., & Baker, J. (2010). Modeling and prediction of chaotic systems with artificial neural networks. *International Journal for Numerical Methods in Fuids, 63.8*, 989–1004.
104. Wu, K. J., & Wang, T. J. (2013). Prediction of chaotic time series based on RBF neural network optimization. *Computer Engineering, 39.10*, 208–216.
105. Wu, X., et al. (2014). Multi-step prediction of time series with random missing data. *Applied Mathematical Modelling, 38.14*, 3512–3522.
106. Xin, B., & Peng, W. (2020). Prediction for chaotic time series-based AE-CNN and transfer learning. *Complexity*, 2680480.
107. Yanan, G., Xiaoqun, C., & Kecheng, P. (2020). Chaotic system prediction using data assimilation and machine learning. In *E3S Web of Conferences* (Vol. 185, p. 02025).
108. Yang, H. Y. et al. (2006). Fuzzy neural very-short-term load forecasting based on chaotic dynamics reconstruction. *Chaos, Solitons and Fractals, 29.2*, 462–469.
109. Yang, F.-P., & Lee, S.-J. (2008). Applying soft computing for forecasting chaotic time series. In *2008 IEEE International Conference on Granular Computing* (pp. 718–723), IEEE.
110. Yeh, J.-P. (2007). Identifying chaotic systems using a fuzzy model coupled with a linear plant. *Chaos, Solitons and Fractals, 32.3*, 1178–1187.
111. Yeo, K. (2019). Data-driven reconstruction of nonlinear dynamics from sparse observation. *Journal of Computational Physics, 395*, 671–689.
112. Yosinski, J. et al. (2014). How transferable are features in deep neural networks? In *Proceedings of the 28th Conference on Neural Information Processing Systems*, 27, 3320–3328.
113. Yu, R., Zheng, S., & Liu, Y. (2017). Learning chaotic dynamics using tensor recurrent neural networks. *Proceedings of the ICML*. In ICML 17 Workshop on Deep Structured Prediction.
114. Yuxia, H., & Hongtao, Z. (2012). Chaos optimization method of SVM parameters selection for chaotic time series forecasting. *Physics Procedia, 25*, 588–594.
115. Zhang, C., et al. (2020). Predicting phase and sensing phase coherence in chaotic systems with machine learning. *Chaos: An Interdisciplinary Journal of Nonlinear Science, 30.8*, 083114.

116. Zhang, J.-S., & Xiao, X.-C. (2000). Predicting chaotic time series using recurrent neural network. *Chinese Physics Letters, 17.2*, 88.
117. Zhang, J., Shu-Hung Chung, H., & Lo, W.-L. (2008). Chaotic time series prediction using a neuro-fuzzy system with time-delay coordinates. *IEEE Transactions on Knowledge and Data Engineering, 20.7*, 956–964.
118. Zhu, Q., Ma, H., & Lin, W. (2019). Detecting unstable periodic orbits based only on time series: When adaptive delayed feedback control meets reservoir computing. *Chaos: An Interdisciplinary Journal of Nonlinear Science, 29.9*, 093125.

Chapter 2
Basic Concepts of Chaos Theory and Nonlinear Time-Series Analysis

Abstract We introduce the basic concepts and methods to formalize and analyze deterministic chaos, with links to fractal geometry. A chaotic dynamic is produced by several kinds of deterministic nonlinear systems. We introduce the class of discrete-time autonomous systems so that an output time series can directly represent data measurements in a real system. The two basic concepts defining chaos are that of attractor—a bounded subset of the state space attracting trajectories that originate in a larger region—and that of sensitivity to initial conditions—the exponential divergence of two nearby trajectories within the attractor. The latter is what makes chaotic dynamics unpredictable beyond a characteristic time scale. This is quantified by the well-known Lyapunov exponents, which measure the average exponential rates of divergence (if positive) or convergence (if negative) of a perturbation of a reference trajectory along independent directions. When a model is not available, an attractor can be estimated in the space of delayed outputs, that is, using a finite moving window on the data time series as state vector along the trajectory.

The concept of chaos, more precisely deterministic chaos, is delicate to define, as it is usually linked to the concept of predictability, such as in the iconic sentence by Edward Lorenz: "Chaos: when the present determines the future, but the approximate present does not approximately determine the future".

In fact, it is much easier to list the properties of a generic chaotic system rather than giving a precise definition of chaos. Roughly speaking, a system is defined as chaotic if:

- Its trajectories remain bounded, meaning that, in the long term, the dynamics are attracted and remain inside a region of the phase space rather than escaping off to infinity (boundedness).
- It is sensitive to initial conditions (sensitivity), i.e., tiny perturbations to the system state become locally amplified in an exponential way, so that two nearby initial conditions soon evolve into very different states, making the system unpredictable in the mid- to long- term. In popular culture, this property is known as the butterfly effect.

© The Author(s), under exclusive license to Springer Nature Switzerland AG 2021
M. Sangiorgio et al., *Deep Learning in Multi-step Prediction of Chaotic Dynamics*,
PoliMI SpringerBriefs,
https://doi.org/10.1007/978-3-030-94482-7_2

These two features produce the aperiodic alternation between expansion and contraction phases of the distance between the system's trajectories [13]. These are the two characterizing forces of chaotic dynamics which take the names of "stretching" and "folding"

In the following section (Sect. 2.1), we introduce the class of dynamical systems that we use as models for real systems and we formalize the first of the above key properties, i.e., the boundedness of the system's trajectories on a so-called attractor. We opt for a description in discrete time only, because it is more directly applicable to data in real systems. Analogous concepts for continuous-time models can be found in standard textbooks, such as [1, 10]. Then, in Sect. 2.2, we formalize the second key property of chaos, the sensitivity to initial conditions, in terms of the well-known Lyapunov exponents. Based on the two concepts of system's attractors and their Lyapunov exponents, in Sect. 2.3 we finally define a chaotic attractor and we discuss its typical fractal geometry. Indeed, chaotic attractors typically present self-similar complex structures and have a non-integer dimension. For this reason, they are also called strange attractors. Section 2.4 extends the two key concepts of attractor and Lyapunov exponents to the case in which a data time series is available, instead of a mathematical model of the system.

2.1 Dynamical Systems and Their Attractors

Let's consider a generic discrete-time system (also called map), whose evolution in one time step is defined by the difference equation:

$$x(t + 1) = f(x(t)), \tag{2.1}$$

where $x(t)$ is the n-dimensional state vector $[x_1(t), x_2(t), \ldots, x_n(t)]^\top$, and $f(\cdot) :$ $\mathbb{R}^n \to \mathbb{R}^n$ is the vector-valued function (or map) whose components $f_1(\cdot)$, $f_2(\cdot)$, $\ldots, f_n(\cdot)$ set the right-hand sides of the state equations in (2.1).

Given an initial condition x_0 at time zero, i.e., $x(0) = x_0$, the sequence of state points $x(t), t = 0, 1, \ldots, T$ is the arc of the system's trajectory, forward in time ($T > 0$), that originates from x_0. If function $f(\cdot)$ is invertible everywhere, the trajectory is uniquely defined also backward in time and the system is said to be reversible.

Note that system (2.1) is autonomous and time-independent, i.e., there is no external input and no time-dependence explicitly affecting function $f(\cdot)$. Such dependencies can be taken into account by letting some of the parameters defining function $f(\cdot)$ be externally driven, as done in Chap. 3.

In the long term, besides the possibility of diverging to infinity, which should never be the case in a realistic model of a real system, the trajectories of system (2.1) typically converge toward a set of state values, which takes the name of attractor. Formally speaking, an attractor is a closed subset A of the state space characterized by the following three properties:

- A is an invariant set under function $f(\cdot)$, i.e., if $x(t)$ is a point of A, so is $x(t+1) = f(x(t))$.
- There exists an open set B, called the basin of attraction for A, which consists of all the initial conditions $x(0)$ that generate trajectories converging forward in time to A.
- A is minimal, in the sense that there is no proper subset of A matching the first two properties.

To be comprehensive, the minimality property is equivalently replaced by some authors by

- the presence of a trajectory densely visiting A;
- or by the following transitivity property: given any two points x' and x'' of A, the evolution of any open set of initial conditions containing x' intersect any open set containing x''.

Attractors can be:

- Fixed points, states $\bar{x}$ such that $f(\bar{x}) = \bar{x}$, corresponding to stationary trajectories.
- Cycles of period $T > 1$, sequence of T states periodically mapped by function $f(\cdot)$.
- Tori (more precisely 1-dim closed curves or d-dim tori, $d > 1$) densely filled by quasi-periodic trajectories.
- Strange attractors, simply speaking attractors not belonging to the above three categories.

Objects of the above four categories can also be present in the system's state space, without being attractors. They can repel nearby trajectories, being called repellers, or attract some and repel others, being called saddles. In particular, the system can have multiple attractors and the boundaries of their attraction basins are typically composed of the initial conditions attracted by saddle objects, their so-called stable manifolds.

Essentially, a strange attractor is the closure of a backbone structure of infinitely many (but countable) saddle equilibria and cycles. Analogously to how the rational numbers (a countable infinite set) densely fill the reals, the backbone of saddle objects and their stable manifolds fill the attractor. Most of the initial conditions in the attractor (technically almost all, i.e., all but a set of zero measure in the state space) do not belong to such objects and manifolds, so that the corresponding trajectories end up visiting (densely) the whole attractor. And so do most of the trajectories that originate in the basin of attraction. We say that this happens "generically" or that it is true for the "generic" initial condition or trajectory, i.e., true with probability one if the initial condition is chosen at random.

Finally, there can also be objects that neither attract nor repel. These are said to be stable, but not asymptotically stable. Their presence is typically due to some simplifying modeling assumptions, such as the absence of friction in mechanics, that introduce the conservation of some energy-like quantity along the system's trajectories.

A complete treatment of the stability of equilibria, cycles, tori, and strange objects is out of the scope of this chapter and the reader is redirected to the above-mentioned standard textbooks [1, 10]. In particular, we will only consider systems with a unique attractor, or limit the analysis to the initial conditions in the basin of the attractor that is relevant for the case under study. Because our aim is to design neural predictors able to perform in the most complex conditions, we will consider the case of chaotic attractors, which are strange attractors characterized by the property of being sensitive to initial conditions. As already anticipated, this property is quantified by the so-called Lyapunov exponents of the attractor, which is the topic of the next section.

2.2 Lyapunov Exponents

The formal way to quantify the extent to which a given trajectory is sensitive to the perturbation of its initial conditions is by its so-called Lyapunov exponents (LEs). Intuitively, in nonlinear systems, the LEs play the same role of the eigenvalues in linear systems.

To compute the LEs associated to a trajectory $x(t), t \geq 0$, taken as reference, it is necessary to define a perturbed trajectory $x(t) + \delta x(t)$ and to linearize the evolution of the perturbation vector $\delta x(t)$ along the reference trajectory $x(t)$. The results is the so-called variational equation:

$$\delta x(t + 1) = J(x(t))\delta x(t), \tag{2.2}$$

where $J(x)$ denotes the Jacobian matrix of function $f(\cdot)$ at x

$$J(x) = \frac{df}{dx} = \begin{bmatrix} \frac{\partial f_1}{\partial x_1} & \cdots & \frac{\partial f_1}{\partial x_n} \\ \vdots & \ddots & \vdots \\ \frac{\partial f_n}{\partial x_1} & \cdots & \frac{\partial f_n}{\partial x_n} \end{bmatrix} \tag{2.3}$$

and the product $J(x(t))\delta x(t)$ is the standard row-column matrix-vector product.

Starting with an initial perturbation $\delta x(0)$ of length ε, the perturbation evolves up to first-order in accordance with the variational equation (2.2) and the perturbed state $x(t) + \delta x(t)$ lies, at time t, on the surface of an ellipsoid (see Fig. 2.1). Changing the initial perturbation $\delta x(0)$ (still of length ε) gives another point $x(t) + \delta x(t)$ on the same ellipsoid's surface. The ellipsoid depends on the initial state $x(0)$ and on the final time t of the trajectory arc. The length ε of the initial perturbation only sets the scale of the ellipsoid due to the linearity of Eq. (2.2).

Up to first-order, the final perturbation $\delta x(t)$ is linearly linked to the initial one by the fundamental solution matrix

$$M(t) = J(x(t-1))J(x(t-2))\ldots J(x(1))J(x(0)) \tag{2.4}$$

Fig. 2.1 First-order approximated evolution of the hypersphere of perturbations $\delta x(0)$ of length ε. After t iterations of the map (2.1), the hypersphere is transformed by the variational equation (2.2) into an (hyper-)ellipsoid, with symmetry axes longer than ε along expanding directions and shorter than ε along contracting directions. The concept is illustrated in two dimensions ($n = 2$)

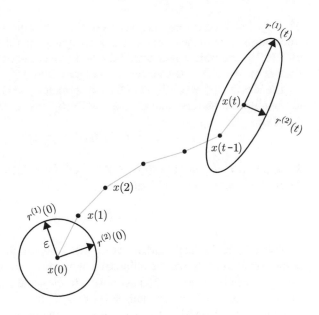

of the variational equation (2.2), i.e.,

$$\delta x(t) = M(t)\delta x(0). \tag{2.5}$$

In other words, the i-th column of $M(t)$ is the solution of (2.2) at time t starting from the natural basis vector $\delta x(0) = v^{(i)}$, with $v_j^{(i)} = 0$ for $j \neq i$ and $v_i^{(i)} = 1$, $i = 1, \ldots n$.

The matrix $M(t)$ is generically nonsingular (it is certainly so for reversible systems) and defines the ellipsoid at time t. Indeed, from

$$\varepsilon^2 = \delta x(0)^\top \delta x(0) = \delta x(t)^\top (M(t)^{-1})^\top M(t)^{-1} \delta x(t)$$
$$= \delta x(t)^\top (M(t)M(t)^\top)^{-1} \delta x(t), \tag{2.6}$$

it follows that the ellipsoid equation

$$\delta x(t)^\top E(t)\delta x(t) = \varepsilon^2 \tag{2.7}$$

holds at any $t \geq 0$ with

$$E(t) = (M(t)M(t)^\top)^{-1}. \tag{2.8}$$

The ellipsoid's symmetry axes are the eigenvectors of matrix $E(t)$. Geometrically, a symmetry axis is a vector r from the ellipsoid center that is aligned with the gradient $E(t)r$ of the quadratic form $\frac{1}{2} r^\top E(t) r$. Denoting by $r^{(i)}(t)$ the symmetry axes associated to the eigenvalue $\lambda_i(t)$ of $E(t)$, the alignment translates into $E(t)r^{(i)}(t) = \lambda_i(t)r^{(i)}(t)$, where $1/\sqrt{\lambda_i(t)} = \sigma_i(t)$ are the so-called singular values of $M(t)$, $i = 1, \ldots n$ (the square roots of the eigenvalues of $E(t)^{-1} = M(t)M(t)^\top$). Recalling that the eigenvectors of a nonsingular matrix coincide with those of the inverse, the ellipsoid's symmetry axes are more commonly referred to as the eigenvectors of $M(t)M(t)^\top$ associated to its eigenvalues $\sigma_i^2(t)$.

The length $r_i(t) = \|r^{(i)}(t)\|$ of the i-th symmetry axis is such that the axis points to the ellipsoid surface, that means that $r^{(i)}(t)$ has to solve the ellipsoid equation (2.7), i.e.,

$$r^{(i)}(t)^\top E(t) r^{(i)}(t) = \varepsilon^2. \tag{2.9}$$

Substituting $r^{(i)}(t) = (1/\sigma_i^2(t)) E(t)^{-1} r^{(i)}(t)$ (the eigenvector equation) in lieu of the second occurrence of $r^{(i)}$ into Eq. (2.9) yields

$$r_i(t) = \varepsilon \sigma_i(t). \tag{2.10}$$

Note that the initial perturbations $r^{(i)}(0)$, $1, \ldots, n$, that are mapped by the variational equation (2.2) into the ellipsoid's symmetry axes at time t (see Fig. 2.1), are the eigenvectors of $M(t)^\top M(t)$ associated to the same singular value $\sigma_i(t)$ of $M(t)$. Indeed, by left-multiplying both sides of $r^{(i)}(0) = M(t)^{-1} r^{(i)}(t)$ by $(M(t)^{-1})^\top$, exploiting at the righthand side the eigen-property of $r^{(i)}(t)$ and Eq. (2.5) to get $(M(t)^{-1})^\top r^{(i)}(0) = (1/\sigma_i^2(t)) M(t) r^{(i)}(0)$, and left-multiplying both sides by $M(t)^\top$, we get

$$r^{(i)}(0) = (1/\sigma_i^2(t)) M(t)^\top M(t) r^{(i)}(0). \tag{2.11}$$

2.2.1 Average Exponents

The Lyapunov exponents (LEs), or Lyapunov characteristic exponents, are, by definition, the log of the average rates of growth of the ellipsoid's symmetry axes, i.e.,

$$L_i = \lim_{t \to +\infty} \log\left(\frac{r_i(t)}{\varepsilon}\right)^{1/t} = \lim_{t \to +\infty} \frac{1}{t} \log \sigma_i(t). \tag{2.12}$$

Indeed, $r_i(t)/\varepsilon = \prod_{k=0}^{t-1} r_i(k+1)/r_i(k)$ is the product of all the t rates of growth of the i-th axis along the reference trajectory and the $(1/t)$-power takes the geometric average (e.g., the average between the rates 2—doubling in one unit of time—and 1/2—halving in one unit of time—is $(2 \times 0.5)^{1/2} = 1$). The limits in (2.12) exist and are finite provided the reference solution $x(t)$ does exist. Note that, by definition, $L_1 \geq L_2 \geq \cdots \geq L_n$, though one does not know a priori the length ordering among the ellipsoid's symmetry axes. Actually, the ordering might change along the trajectory $x(t)$ (especially in the initial phase), so that, in principle, one first computes the limits following the ellipsoid's symmetry axes in arbitrary order and then sort the results (see [15] for a review of the algorithms to numerically compute the LEs). In the following, we will refer to the first symmetry axis $r^{(1)}(t)$ as the longest, to $r^{(2)}(t)$ as the second-longest, etc., i.e., $r_1(t) \geq r_2(t) \geq \cdots \geq r_n(t)$, for sufficiently large t.

The LEs hence measure the average exponential rate of divergence (if positive) or convergence (if negative) of perturbed trajectories with respect to the reference one (the one starting from $x(0)$). Generically, the perturbation $\delta x(t)$ will have a nonzero component along the longest symmetry axes $r^{(1)}(t)$, so that L_1, the largest Lyapunov exponent (LLE), is the dominant rate (average, exponential) of divergence/convergence of the approximate perturbed trajectory (approximated by the variational equation (2.2)). That is, the size of the perturbation grows/decays (on average) as $\varepsilon \exp(L_1 t)$. This is approximately true also for the true perturbed trajectory (the solution of the nonlinear system (2.1) from the perturbed initial condition $x(0) + \delta x(0)$) only if the size ε of the initial perturbation is so small that the perturbed trajectory remains close to the reference one up to time t. However, if $L_1 > 0$, the perturbed trajectory sooner of later leaves the reference one, so that the linearization taken to define the LEs no longer represents the behavior of the perturbation.

The interpretation of the remaining exponents is more delicate. L_2 is the greatest rate chosen from all directions orthogonal to $r^{(1)}(t)$. Similarly, L_3 is the greatest rate chosen from all directions orthogonal to both $r^{(1)}(t)$ and $r^{(2)}(t)$, and so on. However, only those initial perturbations with no component along $r^{(1)}(0)$ (the initial perturbation mapped into $r^{(1)}(t)$ by the variational equation (2.2)) generate (approximate) perturbations at time t orthogonal to $r^{(1)}(t)$. The problem is that considering a different time t modifies the initial perturbation $r^{(1)}(0)$ (it is an eigenvector of the time-dependent matrix $M(t)^{\top}M(t)$), so that the (approximate) perturbation $\delta x(t)$ still grows/decays (on average) as $\varepsilon \exp(L_1 t)$.

The correct interpretation of the non-leading exponents is in terms of partial sums. The sum $L_1 + L_2$ is the dominant rate (average, exponential) for the area of two-dimensional sets of perturbations, as the components along $r^{(3)}(t), \ldots, r^{(n)}(t)$ are dominated (for sufficiently large t) by those along $r^{(1)}(t)$ and $r^{(2)}(t)$ for all the perturbations in the set. Similarly, $\sum_{i=1}^{k} L_i$ is the dominant rate (average, exponential) for the k-dimensional measure (or hypervolume) of k-dimensional sets of perturbations.

Finally, if the trajectory $x(t)$ converge to an attractor A, the resulting limits are generically independent of the initial condition $x(0)$ in the basin of attraction of A. This is certainly true for attracting equilibria, cycles, and tori, whereas it is only generic (i.e., true for almost all initial conditions in the basin of attraction) for strange attractors. In principle, starting from an initial condition on the stable manifold of one of the saddle objects composing the attractor's backbone, the resulting LEs are those characterizing the reached saddle, though unavoidable numerical errors make the trajectory miss the saddle and generically visit the whole attractor.

Exponentially attracting equilibria and cycles have negative LEs, whereas some LEs are null if the speed of attraction is less than exponential (e.g., a time power law); d-dimensional tori have d null exponents, because the internal perturbations are neither expanded nor contracted (on average), the remaining ones being negative in case of exponential attraction; strange attractors typically have at least one positive exponent, responsible of the divergence of the unstable objects in their backbone (this exponent is null if the speed of divergence is weaker than exponential; these attractors are called weakly chaotic). The sum of the attractor's exponents is in any case negative, at least in reversible systems, because n-dimensional volumes of initial conditions in the basin of attractions spread, forward in time, on a lower-dimensional object. As it will be recalled in Sect. 2.3, a chaotic attractor is formally defined as an attractor characterized by $L_1 > 0$.

2.2.2 Local Exponents

The local Lyapunov exponents (loc-LEs) $L_1^{(\text{loc})}(x(t)) \geq L_2^{(\text{loc})}(x(t)) \geq \cdots \geq L_n^{(\text{loc})}(x(t))$ are exponential rates of divergence/convergence faced by the perturbation $\delta x(t)$, locally to the point $x(t)$ of the reference trajectory, along n orthogonal directions $r^{(\text{loc-}i)}(x(t))$ such that $r^{(\text{loc-}1)}(x(t))$ is the dominant one (the direction associated with the fastest expansion if $L_1^{(\text{loc})}(x(t)) > 0$ or slowest contraction if $L_1^{(\text{loc})}(x(t)) < 0$), $r^{(\text{loc-}2)}(x(t))$ is the second-dominant direction (the dominant one among those orthogonal to $r^{(\text{loc-}1)}(x(t))$), etc. Note that the loc-LEs are characteristic of the point $x(t)$ of the trajectory and do not depend on the time t at which the trajectory visits the point.

With reference to a generic point x, the loc-LEs (and corresponding directions) are given by the symmetry axes (lengths and directions) of the ellipsoid formed after one time step by a sphere of perturbation around x, i.e.

$$L_i^{(\text{loc})}(x) = \log \frac{r_i(1)}{\varepsilon} \bigg|_{x(0)=x}. \tag{2.13}$$

From the ellipsoid analysis developed in Subsect. 2.2.1, it follows that the symmetry axes at time 1 are the eigenvectors of $J(x(t)) J(x(t))^\top$. The loc-LEs are therefore the log of the singular values of the system's Jacobian at $x(t)$.

Note that the (average) LEs are not the (arithmetic) time average of the loc-LEs along the reference trajectory. For example, $L_1^{(\text{loc})}(x(t))$ is, at any t, the log of the largest local rate along $r^{(\text{loc-}1)}(x(t))$. In general, however, the longest symmetry axis $r^{(1)}(t)$ of the ellipsoid originated at $x(0)$ is not aligned at time t, with $r^{(\text{loc-}1)}(x(t))$, so that only its component along $r^{(\text{loc-}1)}(x(t))$ is subject to the largest rate. Consequently, the time average of $L_1^{(\text{loc})}(x(t))$ is generically larger than L_1. Similarly, the symmetry axis $r^{(i)}(t)$, $i \geq 2$, is not aligned, at time t, with $r^{(\text{loc-}i)}(x(t))$ and generically has nonzero component along $r^{(\text{loc-}j)}(x(t))$, $j < i$, so that there is not even a clear sign relation between the time average of $L_i^{(\text{loc})}(x(t))$ and L_i.

2.3 Chaotic Systems, Predictability, and Fractal Geometry

2.3.1 Chaotic Attractors and Lyapunov Time Scale

Making use of the concepts defined in the previous sections, we now formally introduce the concept of chaotic attractor, i.e., a closed subset of the state space (invariant under $f(\cdot)$, with its own basin of attraction, and minimal) having at least one positive Lyapunov exponent (i.e., $L_1 > 0$). When there are at least two positive LEs, the attractor is defined as hyperchaotic. A dynamical system with (at least) one chaotic attractor, is said to be chaotic.

The two features, boundedness and sensitivity, presented at the beginning of this chapter to intuitively characterize the chaotic systems, correspond to the formal properties listed above. The sensitivity to initial conditions is guaranteed by the presence of at least one positive LE. A largest Lyapunov exponent (LLE), L_1, greater than 0 is a necessary and sufficient condition to ensure the exponential divergence of any two nearby trajectories ("stretching"). However, if the initial distance between two points is finite (not infinitesimal), the divergence cannot continue indefinitely because of the presence of the attractor, which is bounded by definition. This means that the system nonlinearities will somehow take the two trajectories close again ("folding").

An illustrative representation of stretching and folding is provided in Fig. 2.2.

The formal definition of chaotic attractor reported at the beginning of this subsection allows specifying the distinction between the terms "chaotic" and "strange", which are almost always used interchangeably. An attractor is chaotic if its largest characteristic exponent L_1 is positive. As we have seen, this means that there is an exponential sensitivity to initial conditions. On the other hand, strange attractors do exhibit a sensitive dependence on the initial condition, because of the saddle objects composing their internal structure. However, the divergence of nearby tra-

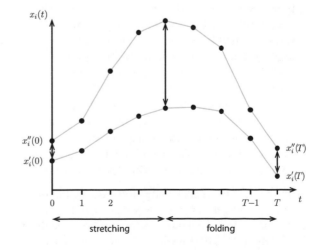

Fig. 2.2 Evolution in time of a generic state variable x_i. The two trajectories, starting from nearby initial conditions, diverge exponentially ("stretching"). After that, they get close together ("folding")

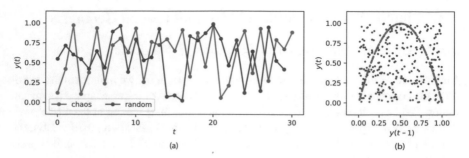

Fig. 2.3 Example of a chaotic (blue) and a random (red) time series. Despite they seem to behave similarly (**a**), the blue dynamic is deterministic as demonstrated by the mapping between $y(t-1)$ and $y(t)$ (**b**)

jectories can be slower than exponential and therefore yield a zero limit in (2.12) for L_1. Hence, chaotic attractors are also strange, while strange attractors can be non-chaotic. Strange non-chaotic attractors are also called "weakly chaotic".

In chaotic regime, the generic system's trajectory shows some peculiar properties:

- Despite its random appearance, a chaotic process is deterministic [2] (see Fig. 2.3), i.e., with the exact same initial conditions, it will always evolve over time in the same way. However, numerical issues (e.g., round-off errors) can generate dynamics that seem to suggest the presence of some source of randomness [12, 21]. For instance, two numerical simulations of the same system, starting from the same initial condition, performed by different hardware architectures, can result in two distinct trajectories visiting the attractor in apparently unrelated ways.
- The trajectories never return to a state already visited (i.e., non-periodicity), but pass arbitrarily close to it. This happens provided that one waits for a sufficient amount of time, because the attractor contains a dense trajectory. In principle, a chaotic system may have sequences of values that exactly repeat themselves (periodic behavior). However, as described in Sect. 2.1, such periodic sequences are repelling rather than attracting, meaning that if the variable is outside the sequence, it will never enter the sequence and in fact, will diverge from it. In this terms, for almost all initial conditions, the variable evolves chaotically with non-periodic behavior. The non-periodicity typical of chaotic dynamics has not to be confused with that of quasi-periodic signals. The difference between the two is usually investigated in the frequency domain. The spectrum of chaotic signals is typically distributed on a range of values, while the one of quasi-periodic signals is concentrated on a finite group of (few) peculiar values.
- The trajectories display complex paths, usually building geometries with a fractal structure: the strange attractors presented in Sect. 2.1.

In the context of time series forecasting, the presence of stretching is the critical factor, even if one has the perfect predictor available. Indeed, the exponential sensitivity to initial condition, soon amplifies the tiniest measurement error or undesired noise on the data feeding the predictor. This is why we often say that "chaotic dynamics are unpredictable". The focus of this book is actually to investigate how far we can go.

Starting from L_1, it is possible to compute the Lyapunov time (LT) of the system, which represents the characteristic time scale on which a dynamical system is chaotic. By convention, the LT is defined as the inverse of L_1, as the time period for the distance between nearby trajectories of the system to increase by a factor of e. The concept of LT is crucial in forecasting tasks, because it mirrors the limits of the predictability of a chaotic system and allows to fairly compare the predictive accuracy in different systems [5, 14, 17, 18, 22].

In accordance with what we have done in Sect. 2.2, it is also possible to define the local Lyapunov time (loc-LT).

2.3.2 Correlation and Lyapunov Dimensions

The geometry of chaotic attractors is usually complex and difficult to describe. It is thus useful to define quantitative properties to characterize these geometrical objects. Dimension is perhaps the most basic property of an attractor. However, the concept of dimension is somewhat ambiguous, since it can have a variety of different definitions [8]. In this subsection, we present two alternatives for measuring the attractor dimensions: the correlation and Lyapunov dimensions.

The correlation dimension is a fractal dimension specifically designed for the attractors of a dynamical system. Consider a target system's attractor A and let $\{x(t),\ t = 0, \ldots, T\}$ be an arc of trajectory on the attractor A. For any given radius $r > 0$, define the correlation function

$$C(r, T) = \frac{\#\,\text{pairs}\,(x(t'), x(t''))\ :\ \|x(t') - x(t'')\| < r,\ 0 \le t' < t'' \le T}{T(T + 1)/2}, \quad (2.14)$$

where $T(T + 1)/2$ is the number of point pairs in the trajectory arc. The correlation function increases from 0 to 1 as r increases from 0 to the maximum distance between the points in the trajectory. Let d, possibly non-integer, be the attractor's dimension to be determined. For small r, the number of points $x(t)$ that are r-close to a given point $x \in A$ increases linearly with the length T of the trajectory arc and proportionally to r^d. Think, for example, to the case $d = 1$ (a one-dim closed curve densely filled by a quasi-periodic trajectory). The segment of the curve contained in the r-ball centered at x has a length proportional to r^d and is filled by the trajectory as T goes to infinity. For large T and small r, the numerator of the correlation function is hence proportional to $T^2 r^d$ (there are T points each with a number of r-close neighbors proportional to $T r^d$), while the denominator has order T^2 (see (2.14)). Solving for d gives the correlation dimension

$$d_{\text{corr}}(A) = \lim_{r \to 0} \lim_{T \to \infty} \frac{\log C(r, T)}{\log r}. \quad (2.15)$$

Fig. 2.4 Schematic illustration of the estimation procedure of the correlation dimension. The estimate is obtained as $d_{corr}(A) = \Delta \log C(r)/\Delta \log r$

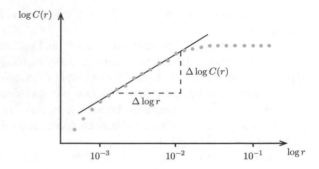

In practice, we work with finite r and T. Starting from an initial condition $x(0)$ in the basin of attraction of the target attractor, we discard the initial trajectory transient (say, the first t_0 points) and then take $T + 1$ points $x(t_0), x(t_0 + 1), \ldots, x(t_0 + T)$, for sufficiently large t_0 and T. We then compute the correlation function for a sequence of small increasing values of the radius r. Plotting $\log C(r)$ versus $\log r$, there will be an interval of r in which the obtained points will approximately align. The correlation dimension $d_{corr}(A)$ is given by the slope of this linear part of the graph (see, Fig. 2.4).

To well identify the linear part of the graph, it is worth to discuss how the graph gets distorted for too small and too large values of the radius r. The behavior for small r is due to the finite resolution with which our finite arc of trajectory fills the attractor. When r is smaller than such a resolution, the correlation function underestimates the density of the attractor points within distance r and scaling this underestimate with r^d necessarily results in a d larger than the actual attractor's dimension. This is evident in the expression (2.15) by considering smaller and smaller r at constant T: $C(r, T)$ vanishes below a resolution threshold on r, so that the numerator diverges to $-$ infinity, while the denominator is still finite. Vice-versa, for too large values of r, the boundedness of the attractor makes the correlation function saturate at 1 (independently on how large is T) and the slope of the graph correspondingly vanishes (see again Fig. 2.4).

When the full spectrum of the attractor's LEs is available, a simple way of estimating the attractor's dimension is the well-known Kaplan-Yorke formula [9], which approximates the concept of Lyapunov dimension. The geometric idea behind the concept is related to the interpretation given in Sunsect. 2.2.1 of the partial sums of the attractor's LEs. According to that interpretation, the Lyapunov dimension $d_{Lyap}(A)$ is the dimension of a small set of initial conditions, in the basin of attraction, with $d_{Lyap}(A)$-dimensional measure (hypervolume) that neither grows nor decays (on average), as the corresponding trajectories converge to the attractor. Because the trajectories densely visit the attractor, this occurs when the set of initial conditions has the same dimension of the attractor.

The dimension could however be non-integer, so that the best one can do is to look for the index k for which $\sum_{i=1}^{k} L_i \geq 0$ and $\sum_{i=1}^{k+1} L_i < 0$. The hypervolume of a k-dimensional set of initial conditions grows (on average), while the one of a $(k + 1)$-dimensional set does vanish, while converging to the attractor, so that all we

Fig. 2.5 Schematic illustration of the Kaplan-Yorke estimation of the fractal dimension

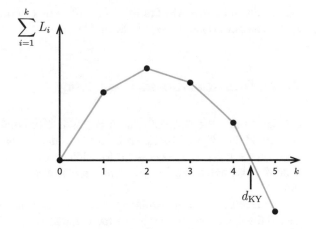

can say is that the dimension d is in between k and $k + 1$. As illustrated in Fig. 2.5, the Kaplan-Yorke formula simply considers a continuous piece-wise interpolation of the partial sums of the attractor's LEs. The Lyapunov dimension $d_{\text{Lyap}}(A)$ is therefore approximated with the value at which the piece-wise interpolation crosses the horizontal axis, the latter representing the zero-rate of expansion/contraction. The result is the following formula:

$$d_{\text{KY}}(A) = k + \frac{\sum_{i=1}^{k} L_i}{|L_{k+1}|}. \tag{2.16}$$

To conclude this subsection on attractors dimension, we clarify the relationship between fractals and strange attractors. So far, we described these latter as complex geometrical objects which usually have a non-integer dimension. In fact, these two features (being strange and having a fractal structure) are generally strongly related, but not exactly equivalent. The attractors characterized by a non-integer dimension are always strange. Conversely, it is possible to find some peculiar cases where a strange attractor has integer dimension. A classical example is the logistic map with a parameter (i.e., the growth rate) equal to 4 (the system is presented in Subsect 3.1.1). In such a case, it can be proved that the chaotic orbit is dense in the unit interval [0, 1], and thus its dimension is equal to 1 (see [1], page 235).

2.4 Attractor Reconstruction from Data

The concepts presented in the first part of this chapter require to know the state space representation of the dynamical system. In practical applications, the equations describing the evolution of the state variables (and also the whole state vector itself) are usually unknown. Most of the time, only a time series of observed values of the system output $y(t) = g(x(t))$, is available.

2.4.1 Delay-Coordinate Embedding

Starting from a sequence of data, it is possible to build a m-dimensional delayed version of the original phase space by considering the delay coordinates $Y(t) = [y(t), y(t-1), \ldots, y(t-m+1)]$ for each time t in the time series (Fig. 2.3). This procedure is known as delay-coordinate reconstruction, or delay-coordinate embedding.

The reconstruction is extremely useful because, under broad conditions, it preserves the dynamical properties of the original system, including, for instance, its asymptotic behavior and the LEs (the first k, being k the index of the Kaplan-Yorke formula, see Subsect. 2.4.2 for further details).

More specifically, the Takens' theorem [20] demonstrates that a smooth attractor can be reconstructed from a univariate sequence of observations made with a generic function $g(\cdot)$ if the embedding dimension, m, is large enough. Hence, the theorem forms a bridge between the theory of nonlinear dynamical systems and the analysis of time series.

The reconstructed attractor, which lies in the m-dimensional delay-coordinate space, is topologically equivalent to the original attractor (in the initial n-dimensional state space). In other words, the reconstructed dynamic, i.e., $Y(t+1) = F(Y(t))$, is equivalent to that of the original state space; more exactly, they are related by a smooth, invertible change of coordinates (i.e., a diffeomorphism). Thanks to these properties of the mapping between the original and the reconstructed spaces, every state in the phase space of the system can be uniquely represented by the output data. The reconstructing procedure, however, does not preserve the geometric shape of the structures of the original attractor.

Under a geometrical perspective, the idea behind Takens' theorem is to unfold the reconstruction of the trajectories sufficiently to avoid self-crossing. Trajectories cannot cross because a deterministic dynamical system's evolution is fully defined by its current state. This means that starting from a given point in phase space, the next point is uniquely determined. As a consequence, two trajectories meeting at the same point in the phase space cannot evolve in two different ways.

In practice, due the numerical issues and the possible presence of noise, one would verify that, in a space with a sufficient embedding dimension, two nearby points should be mapped by a single iteration of the function $F(\cdot)$ in relatively close positions. If this is not the case, i.e., the points evolve in completely different positions after one time step, this is probably due to the fact that we are in presence of self-crossings caused by an insufficient embedding dimension. For large enough m, these points will move away from each other. This situation, however, can occur also when using an appropriate value of m, if $F(\cdot)$ is highly irregular and has steep valleys and ridges. It is a fairly uncommon case because deterministic chaos is usually characterized by simple maps, which produce unexpectedly complex dynamics when applied iteratively, but its occurrence cannot be excluded a priori.

The practical way usually adopted to estimate the minimal embedding dimension from data is the so-called false nearest neighbor algorithm [11]. The idea behind this algorithm is to examine how the number of neighbors of a point in the reconstructed space varies with increasing embedding dimension. When the embedding dimension is too low, many of the points that are in reality far apart could seem to be neighbors as a consequence of projecting into a space of smaller dimension. These points are called false neighbors. In an appropriate (or higher) embedding dimension, the false neighbors will no longer be neighbors. Conversely, two points that are near to each other in the sufficient embedding dimension should remain close as the dimension increases.

A visual representation of the concept of true and false neighbor is reported in Fig. 2.6. In a 1-dimensional embedding, the three considered points—denoted with **A**, **B**, and **C**—seem to be neighbors. Increasing the embedding dimension by one, we discover that **A** and **B** are still neighbors, while **B** and **C** are actually false neighbors. Embedding the system in three dimensions, one would see that the points **A** and **B** remain close, although **C** is still far from **B**. If the situation described above occurs for

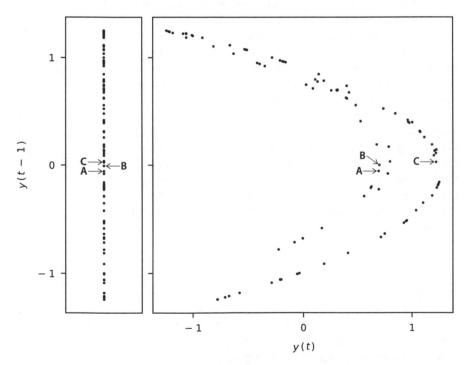

Fig. 2.6 1- and 2-dimensional embeddings of a noise-free dataset obtained simulating the Hénon map (presented in Sect. 3.1), with $y(t) = x_1(t)$. The points **A** and **B** are true neighbors, while **B** and **C** are false neighbors

Fig. 2.7 Fraction of false nearest neighbors computed for an increasing embedding dimension (m)

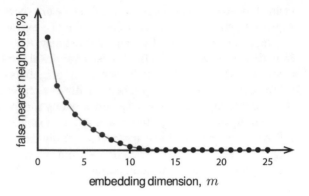

all the points in the dataset, we will conclude that in this specific case, a 2-dimensional space is sufficient to unfold the attractor.

Practically speaking, two points are considered as neighbors when their Euclidean distance is lower than a certain threshold, that has to be properly selected depending on the specific application. The algorithm checks the neighbors in increasing embedding dimensions until it finds only a negligible number of false neighbors Fig. 2.7. This is chosen as the lowest embedding dimension, which is presumed to give reconstruction without self-intersections between the trajectories.

If the data are noise-free, the percentage of false neighbors will drop to zero when the proper dimension is reached. In the case of noisy time series, we expect that a certain (hopefully small) number of false neighbors remain also for large embedding dimension due to the presence of noise.

Takens' theorem also states that, generically, the m has to be only a little larger, more precisely, the first integer larger, than twice the dimension of the attractor. To develop a geometric intuition about this fact, we can consider the following simple example. Two 1-dimensional curves can have intersections in either $\mathbb{R}^2$, $\mathbb{R}^3$ or any higher-dimensional space. The difference is that a small perturbation will remove the intersection in $\mathbb{R}^3$, while in $\mathbb{R}^2$ it will only move the intersection somewhere else. We can imagine that the two curves correspond to two invariant strands of the attractor, e.g., two arcs belonging to the reconstruction of a 1-dimensional torus with a quasi-periodic motion on it. A slight perturbation in the system dynamics would generically cause the self-intersection in $\mathbb{R}^3$ to disappear.

In other words, although it is certainly possible for two generic curves placed at random to have intersections in $\mathbb{R}^k$ for $k \geq 3$, we should consider them as exceptional situations and expect them to occur with essentially zero probability. This is the reason why we used the term "generically" hereinbefore: if $m > 2d$, self-crossings should in principle be rare, but they could still occur. When this is the case, it is necessary to further increase m, even if it is already greater than $2d$.

2.4.2 Estimation of the Largest Lyapunov Exponent

As reported in Sect. 2.2, the traditional algorithm for the computation of the LEs makes use of the Jacobian matrix and thus requires knowledge of the state equations of the system. However, when the model of the system is not available, it is necessary to make use of statistical methods for detecting chaos [7].

Once an appropriate embedding dimension m of the dataset has been estimated, the dynamic of the system can be analyzed in the delayed phase space $Y(t) = [y(t), y(t-1), \ldots, y(t-m+1)]$ which shares the same topological properties (and thus also L_1) of the original phase space as stated by Takens' theorem [20]. After that, the following procedure [23] can be performed to estimate L_1:

1. select N_p pairs of nearby points in the m-dimensional delayed space, $Y(t_i)$ and $Y(t_j)$;
2. for each pair, compute the Euclidean distance between the two points $\delta_p(0) = \left\| Y(t_i) - Y(t_j) \right\|$;
3. recompute the distance between the two points of each pair after t_e time steps $\delta_p(t_e) = \left\| Y(t_i + t_e) - Y(t_j + t_e) \right\|$. On average, for small values of t_e, we expect that $\delta_p(t_e)$ evolves following the exponential law:

$$\delta_p(t_e) = \delta_p(0) \cdot \exp(L_1 t_e); \tag{2.17}$$

4. calculate $Q(t_e)$, namely the average logarithm of the divergence rate as:

$$Q(t_e) = \frac{1}{N_p} \sum_{p=1}^{N_p} \left(\log \frac{\delta_p(t_e)}{\delta_p(0)} \right). \tag{2.18}$$

Repeating the procedure for increasing values of the expansion step t_e, we can plot $Q(t_e)$ as a function of t_e. In a chaotic system, the initial part of this curve is characterized by a linear trend (see Fig. 2.8). This is due to the fact that two points initially close together will diverge exponentially, and that this exponential expansion produces a linear trend because of the logarithm. After that, the average divergence tends to a constant value because the trajectories lie inside the chaotic attractor. Finally, the L_1 can be easily computed as the slope of the function $Q(t_e)$ in the initial part.

The algorithm presented in this subsection only provides an estimation of L_1. Alternative procedures for computing the largest exponent are available [6, 16, 24]. In principle, one could also make use of more complex procedures which allow us to numerically compute the positive Lyapunov spectrum, i.e., only positive exponents [23], or the whole spectrum of LEs [3, 4, 19]. However, even in this last case, only the estimates of the first k exponents make sense. Going further would produce arbitrary values since the data are typically sampled when the trajectories are already in the attractor, meaning that they are representative of what happens inside it, while the LEs beyond the k-th ($k+1, k+2, \ldots$) describe the transient of convergence toward

Fig. 2.8 Schematic illustration of the estimation procedure of the LLE from data. The estimate is obtained as $L_1 = \Delta Q(t_e)/\Delta t_e$

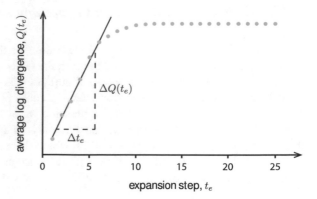

the attractor. In any case, the dimension of the original state space remains unknown, thus we do not even know how many LEs should be computed.

In this book, we limit the analysis to L_1 because the computation of the LEs spectrum presents computational issues which strongly affect the numerical stability of the estimation [23]. Note, however, that in the context of time series prediction, we are interested in deriving the Lyapunov time LT, which sets the limit of predictability of the chaotic system. To this end, it is sufficient to know L_1, since it defines the system's dominant dynamics.

References

1. Alligood, K. T., Sauer, T. D., & Yorke, J. A. (1996). *Chaos*. Springer.
2. Boeing, G. (2016). Visual analysis of nonlinear dynamical systems: Chaos, fractals, self-similarity and the limits of prediction. *Systems, 4.4*, 37.
3. Briggs, K. (1990). An improved method for estimating Liapunov exponents of chaotic time series. *Physics Letters A, 151.1-2*, 27–32.
4. Brown, R., Bryant, P., & Abarbanel, H. D. (1991). Computing the Lyapunov spectrum of a dynamical system from an observed time series. *Physical Review A, 43.6*, 2787.
5. Dercole, F., Sangiorgio, M., & Schmirander, Y. (2020). An empirical assessment of the universality of ANNs to predict oscillatory time series. *IFAC-PapersOnLine, 53.2*, 1255–1260.
6. Ellner, S., et al. (1991). Convergence rates and data requirements for Jacobian-based estimates of Lyapunov exponents from data. *Physics Letters A, 153.6-7*, 357–363.
7. Ellner, S., & Turchin, P. (1995). Chaos in a noisy world: New methods and evidence from time-series analysis. *The American Naturalist, 145.3*, 343–375.
8. Farmer, J. D., Ott, E., & Yorke, J. A. (1983). The dimension of chaotic attractors. *Physica D: Nonlinear Phenomena, 7.1-3*, 153–180.
9. Frederickson, P., et al. (1983). The Liapunov dimension of strange attractors. *Journal of Differential Equations, 49.2*, 185–207.
10. Guckenheimer, J., & Holmes, P. (2013). Nonlinear oscillations, dynamical systems, and bifurcations of vector fields. (Vol. 42). Springer Science and Business Media.
11. Kennel, M. B., Brown, R., & Abarbanel, H. D. (1992). Determining embedding dimension for phase-space reconstruction using a geometrical construction. *Physical Review A, 45.6*, 3403.

12. Nepomuceno, E. G., et al. (2019). Soft computing simulations of chaotic systems. *International Journal of Bifurcation and Chaos, 29.08*, 1950112.
13. Ott, E. (2002). *Chaos in dynamical systems*. Cambridge University Press.
14. Pathak, J., et al. (2018). Model-free prediction of large spatiotemporally chaotic systems from data: A reservoir computing approach. *Physical Review Letters, 120.2*, 024102.
15. Ramasubramanian, K., & Sriram, M. S. (2000). A comparative study of computation of Lyapunov spectra with different algorithms. *Physica D: Nonlinear Phenomena, 139.1-2*, 72–86.
16. Rosenstein, M. T., Collins, J. J., & De Luca, C. J. (1993). A practical method for calculating largest Lyapunov exponents from small data sets. *Physica D: Nonlinear Phenomena, 65.1-2*, 117–134.
17. Sangiorgio, M. (2021). Deep learning in multi-step forecasting of chaotic dynamics. Ph.D. thesis. Department of Electronics, Information and Bioengineering, Politecnico di Milano.
18. Sangiorgio, M., & Dercole, F. (2020). Robustness of LSTM neural networks for multi-step forecasting of chaotic time series. *Chaos, Solitons and Fractals, 139*, 110045.
19. Sano, M., & Sawada, Y. (1985). Measurement of the Lyapunov spectrum from a chaotic time series. *Physical Review Letters, 55.10*, 1082.
20. Takens, F. (1981). Detecting strange attractors in turbulence. In: *Dynamical systems and turbulence, Warwick 1980* (pp. 366–381). Springer.
21. Ushio, T., & Hsu, C. (1987). Chaotic rounding error in digital control systems. *IEEE Transactions on Circuits and Systems, 34.2*, 133–139.
22. Vlachas, P. R., et al. (2020). Backpropagation algorithms and Reservoir Computing in Recurrent Neural Networks for the forecasting of complex spatiotemporal dynamics. *Neural Networks, 126*, 191–217.
23. Wolf, A., et al. (1985). Determining Lyapunov exponents from a time series. *Physica D: Nonlinear Phenomena, 16.3*, 285–317.
24. Wright, J. (1984). Method for calculating a Lyapunov exponent. *Physical Review A, 29.5*, 2924.

Chapter 3
Artificial and Real-World Chaotic Oscillators

Abstract Four archetypal chaotic maps are used to generate the noise-free synthetic datasets for the forecasting task: the logistic and the Hénon maps, which are the prototypes of chaos in non-reversible and reversible systems, respectively, and two generalized Hénon maps, which represent cases of low- and high-dimensional hyperchaos. We also present a modified version of the traditional logistic map, introducing a slow periodic dynamic of the growth rate parameter, that includes ranges for which the map is chaotic. The resulting system exhibits concurrent slow and fast dynamics and its forecasting represents a challenging task. Lastly, we consider two real-world time series of solar irradiance and ozone concentration, measured at two stations in Northern Italy. These dynamics are shown to be chaotic movements by means of the tools of nonlinear time-series analysis.

In the past, many nonlinear systems that exhibit a chaotic behavior have been discovered in several fields of science, from ecology [20] to finance [17, 18], from population dynamics [10] to atmospheric processes [13].

3.1 Artificial Chaotic Systems

In this book, we consider four discrete-time dynamical systems universally known for being chaotic to test the predictive power of the neural predictors: the logistic map, the Hénon map, and two generalized Hénon maps. The logistic and Hénon maps are the classical prototypes of chaos in non-reversible and reversible discrete-time systems, respectively, whereas the generalized Hénon map is considered to include low- and high-dimensional hyperchaos. We limit the analysis to a single output variable for each system so that the corresponding time series is univariate. For the ease of the reader, the systems are presented below.

© The Author(s), under exclusive license to Springer Nature Switzerland AG 2021
M. Sangiorgio et al., *Deep Learning in Multi-step Prediction of Chaotic Dynamics*,
PoliMI SpringerBriefs,
https://doi.org/10.1007/978-3-030-94482-7_3

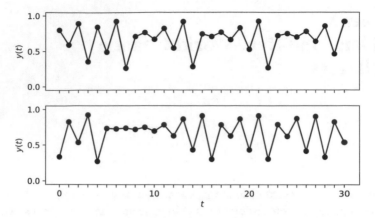

Fig. 3.1 Two examples of the logistic map behavior ($r = 3.7$) when starting from different initial conditions selected at random

3.1.1 Logistic Map

The logistic map is a one-dimensional quadratic map used to describe biomass or economic growth. It has only one state variable, the density of the biological or economic resource that we take equal to the output, so that the system can be defined by a single difference equation that directly describes the dynamics of the output, i.e.,

$$y(t + 1) = r \cdot y(t) \cdot \left(1 - y(t)\right), \tag{3.1}$$

where $r > 1$ is the growth rate at low density. As is well known, the logistic map exhibits chaotic behavior for most of the values of r in the range 3.6-4. Two examples of how the logistic map ($r = 3.7$) variable evolves in time are reported in Fig. 3.1.

The equation describing the map is a simple quadratic polynomial, which assumes the shape represented in the first panel of Fig. 3.2. Reproducing this function with a neural net is an easy task. Things start to become more and more complex when this simple map is iterated (other panels in Fig. 3.2). This is a general property of chaotic systems that can immediately be seen in the logistic map. The effect of the increasing complexity that occurs when the map is iterated is that small errors in the input (on the horizontal axis in Fig. 3.2) give larger and larger errors in the output (vertical axis). A mid- to long-term prediction of chaotic systems is, therefore, intrinsically problematic, even if one knows the true model generating the data [2, 3, 15].

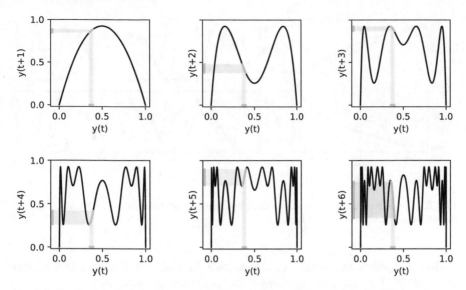

Fig. 3.2 Graph of the logistic map ($r = 3.7$, first panel on the left) and of the 2-to-6 iterated maps. Grey areas show how a small error at time t propagates when the map is iterated

3.1.2   Hénon Map

The Hénon map, a toy model of celestial mechanics, is a two-dimensional ($n = 2$) system traditionally defined in state space [6]:

$$\begin{cases} x_1(t + 1) = 1 - a \cdot x_1(t)^2 + x_2(t) \\ x_2(t + 1) = b \cdot x_1(t), \end{cases} \tag{3.2}$$

where a and b are model parameters, usually taking values 1.4 and 0.3, respectively. Considering the first state variable as output, $y(t) = x_1(t)$, the system is equivalent to the following two-dimensional ($m = 2$) nonlinear regression:

$$y(t + 1) = 1 - a \cdot y(t)^2 + b \cdot y(t - 1). \tag{3.3}$$

The Hénon attractor and the evolution in time of the output variable $y(t)$ are represented in Fig. 3.3.

3.1.3   Generalized Hénon Map

The generalized Hénon map is an extension of the Hénon map introduced with the purpose of generating hyperchaos (a chaotic attractor characterized by at least two

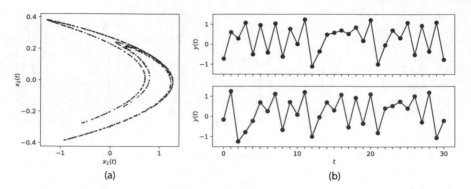

Fig. 3.3 Chaotic attractor produced by the Hénon map (**a**), and evolution in time of the system output for two random initial conditions (**b**)

positive Lyapunov exponents, i.e., at least two directions of divergence within the attractor) [1, 11]. The state equations of the n-dimensional case are:

$$\begin{cases} x_1(t+1) = a - x_{n-1}(t)^2 - b \cdot x_n(t) \\ x_j(t+1) = x_{j-1}(t), \quad j = 2, \dots, n. \end{cases} \tag{3.4}$$

(for $n = 2$, using $x_1(t)/a$ and $(-b/a)x_2(t)$ as new coordinates and $-b$ as a new parameter gives the traditional formulation (3.2)). Hyperchaotic behavior is observed for $a = 1.9$ and $b = 0.03$. The system can be rewritten as a nonlinear regression on the output, $y(t) = x_1(t)$, obtaining:

$$y(t+1) = a - y(t - m + 2)^2 - b \cdot y(t - m + 1). \tag{3.5}$$

We consider the 3D and the 10D generalized Hénon maps. The first is characterized by $n = m = 3$. The corresponding Lyapunov exponents are $L_1 = 0.276$, $L_2 = 0.257$, and $L_3 = -4.040$. The attractor's fractal dimension, computed with the Kaplan-Yorke formula [4], is 2.13. Figure 3.4 shows the chaotic attractor and the evolution in time of the system's output starting from two different initial states. The second ($n = m = 10$) has 9 positive Lyapunov exponents and its attractor's fractal dimension is equal to 9.13.

3.1.4   Time-Varying Logistic Map

As a further test case, we consider a non-stationary system with both slow and fast dynamics. These kinds of processes characterize many natural phenomena and for this reason attract the attention of many researchers in the field of dynamical systems theory (see, for instance, [12, 19]).

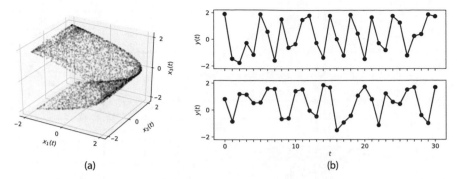

(a)   (b)

Fig. 3.4 Chaotic attractor produced by the 3D generalized Hénon map (**a**), and evolution in time of the system output for different initial conditions selected at random (**b**)

The slow-fast system presented here is obtained introducing, in the traditional logistic map, a slow (periodic) dynamic for the parameter $r(t)$ [14]:

$$\begin{cases} y(t+1) = r(t) \cdot y(t) \cdot \left(1 - y(t)\right) \\ r(t) = \dfrac{r_{max} + r_{min}}{2} + \dfrac{r_{max} - r_{min}}{2} \cdot \sin(\dfrac{t \cdot \pi}{5000}), \end{cases} \tag{3.6}$$

where $r_{min} = 2.9$ and $r_{max} = 3.7$ represent the lower and upper bound of the growth rate. These values have been selected so that the system in turn exhibits stable, periodic or chaotic behavior, as shown in Fig. 3.5.

Testing the neural predictors presented in Sect. 4.1 on this kind of system is interesting because the forecasting task requires to retain information about both the slow-varying context (long-term memory), and the fast dynamics of the logistic map. Moreover, this task has different degrees of complexity: it is fairly simple when the system has stable or periodic behavior, and much more complex when the dynamic becomes chaotic.

Formally speaking, the system defined by Eqs. (3.6) meets all the conditions listed in Chap. 2, and thus it is a chaotic system.

3.2 Real-World Time Series

3.2.1 Solar Irradiance

As anticipated at the beginning of this chapter, chaos theory does not only consider analytical systems like those presented above; a remarkable variety of natural phenomena are thought to be chaotic, from meteorology to the orbits of celestial bodies, from chemical reactions to electronic circuits [21].

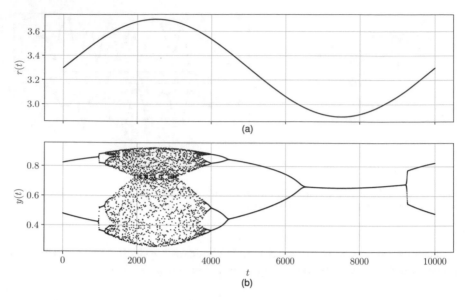

Fig. 3.5 Dynamics of the parameter $r(t)$ (**a**) and of the logistic map variable $y(t)$ (**b**) for 10,000 steps

The first real-world dataset considered in this study is the solar irradiance recorded from 2014 to 2019 by a Davis Vantage 2 weather station installed and managed by the Politecnico di Milano at Como Campus, Northern Italy. The station was continuously monitored and checked for consistency as part of the dense measurement network of the Centro Meteorologico Lombardo (www.centrometeolombardo.com). Its geographic coordinates are: Lat = 45.80079, Lon = 9.08065 and Elevation = 215 m asl. Together with the solar irradiance, the following physical variables are recorded every 5 min: air temperature, relative humidity, wind speed and direction, atmospheric pressure, rain, and the UV index. However, the current study only adopts purely autoregressive models. A detail of the time series, recorded at each hour in Como, is shown in Fig. 3.6a.

We can interpret this time series as the sum of three different components: the astronomical condition (namely the position of the sun), which produces the evident annual and daily cycles; the current meteorological situation (the attenuation due to atmosphere, including clouds); and the specific position of the receptor that may be shadowed by the passage of clouds in the direction of the sun. The first component is deterministically known, the second can be predicted with a certain accuracy, while the third is much trickier and may easily vary within minutes without a clear dynamic. The expected global solar radiation in average clear sky conditions (see Fig. 3.6b) was computed by using the Ineichen and Perez model, as presented in [7, 9]. The Python code that implements this model is part of the SNL PVLib Toolbox, provided by the Sandia National Labs PV Modeling Collaborative (PVMC) platform [16].

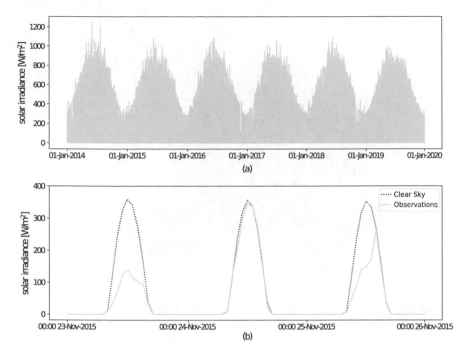

Fig. 3.6 Hourly solar irradiance time series (**a**), compared with clear sky values for a few specific days (**b**)

Three additional solar irradiance time series are considered to evaluate the generalization capability of the predictors [5]: Casatenovo (year 2011), Bigarello (2016), and Bema (2017). In order to provide heterogeneous datasets, the data are recorded in different sites spanning more than one degree of latitude and representing quite different geographical settings: from the low and open plain at 35 m.a.s.l to the mountains at 800 m.a.s.l (Fig. 3.7).

3.2.2 Ozone Concentration

The second real-world case study considered is the ozone (O_3) concentration time series recorded from 2008 to 2017 in Chiavenna, Northern Italy (the station's geographic coordinates are: Lat = 46.32080, Lon = 9.39559 and Elevation = 333 m asl). The dataset is maintained and made available by the regional agency for environment (ARPA in the Italian acronym) of the Lombardy region (www.arpalombardia.it).

Ground level ozone formation is known to be a complex phenomenon. Ozone is a secondary pollutant: there are no ozone emissions, but its presence is due to chemical reactions involving other pollutants (called precursors). These photochemical reactions are activated by the emission of nitrogen oxides (NO_X), mainly due to road

Fig. 3.7 Map of the sites where the solar irradiance datasets have been recorded. The Lombardy region territory, in Northern Italy, is represented in grey. The main dataset is recorded in Como (red marker). The three additional datasets are recorded in correspondence with the green markers

transportation and domestic heating, and volatile organic compounds (VOC) due to industrial plants, road transportation and agriculture. The spatial distribution of the ozone precursors is reported in Fig. 3.8a. The precursors are not sufficient to complete the photochemical reactions, ultraviolet radiation is also needed. The process briefly mentioned above takes a few hours, and thus the ozone concentration may reach high values kilometers away form the precursors' sources, depending on the meteorological conditions (see Fig. 3.8b).

The tropospheric ozone formation is known to be a nonlinear process [8], and thus represents an interesting case study to test the neural predictors presented in this book. A detail of the hourly time series recorded in Chiavenna is shown in Fig. 3.9.

Like almost all the environmental variables, the ozone concentration also has daily and annual periodic trends caused by the Earth's rotation on its axis and revolution around the Sun (Fig. 3.9). However, as shown in Fig. 3.9b, in this case, the daily cycle is less evident than in the solar irradiance dataset (see Fig. 3.6b).

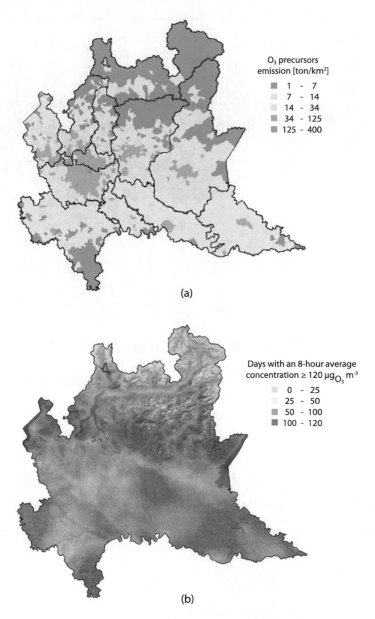

(a)

(b)

Fig. 3.8 Ozone precursors emissions in 2014 (**a**), and number of days with an 8-hour average concentration exceeding 120 $\mu g/m^3$ in 2017 (**b**) in the Lombardy region. The boundary of Chiavenna municipality is shown in red. Images from ARPA Lombardia (www.arpalombardia.it)

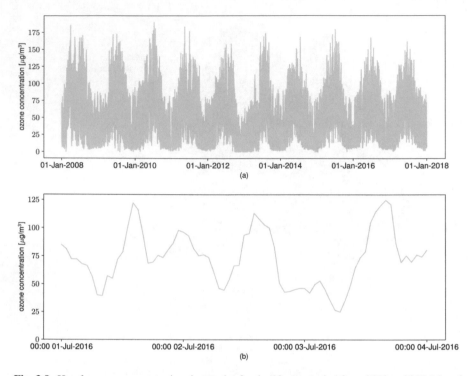

Fig. 3.9 Hourly ozone concentration time series for the 10-year period from 2008 to 2017 (**a**) and for a few specific days (**b**)

References

1. Baier, G., & Klein, M. (1990). Maximum hyperchaos in generalized Hénon maps. *Physics Letters A, 151.6-7*, 281–284.
2. Bollt, E. (2020). Regularized kernel machine learning for data driven forecasting of chaos. *Annual Review of Chaos Theory, Bifurcations and Dynamical Systems, 9*, 1–26.
3. Dercole, F., Sangiorgio, M., & Schmirander, Y. (2020). An empirical assessment of the universality of ANNs to predict oscillatory time series. *IFAC-PapersOnLine, 53.2*, 1255–1260.
4. Frederickson, P., et al. (1983). The Liapunov dimension of strange attractors. *Journal of Differential Equations, 49.2* , 185–207.
5. Guariso, G., Nunnari, G., & Sangiorgio, M. (2020). Multi-step solar irradiance forecasting and domain adaptation of deep neural networks. *Energies, 13.15*, 3987.
6. Hénon, M. (1976). A two-dimensional mapping with a strange attractor. In *The Theory of Chaotic Attractors* (pp. 94–102). Springer.
7. Ineichen, P., & Perez, R. (2002). A new airmass independent formulation for the Linke turbidity coefficient. *Solar Energy, 73.3*, 151–157.
8. Lin, X., Trainer, M., & Liu, S. C. (1988). On the nonlinearity of the tropospheric ozone production. *Journal of Geophysical Research: Atmospheres, 93.D12*, 15879–15888.
9. Perez, R., et al. (2002). A new operational model for satellite-derived irradiances: Description and validation. *Solar Energy, 73.5*, 307–317.
10. Rai, V., & Upadhyay, R. K. (2004). Chaotic population dynamics and biology of the top-predator. *Chaos, Solitons and Fractals, 21.5*, 1195–1204.

11. Richter, H. (2002). The generalized Henon maps: Examples for higher-dimensional chaos. *International Journal of Bifurcation and Chaos, 12.06*, 1371–1384.
12. Rossetto, B., et al. (1998). Slow-fast autonomous dynamical systems. *International Journal of Bifurcation and Chaos, 8.11*, 2135–2145.
13. Russell, F. P., et al. (2017). Exploiting the chaotic behaviour of atmospheric models with reconfigurable architectures. *Computer Physics Communications, 221*, 160–173.
14. Sangiorgio, M. (2021). Deep learning in multi-step forecasting of chaotic dynamics. Ph.D. thesis. Department of Electronics, Information and Bioengineering, Politecnico di Milano.
15. Sangiorgio, M., & Dercole, F. (2020). Robustness of LSTM neural networks for multi-step forecasting of chaotic time series. *Chaos, Solitons and Fractals, 139*, 110045.
16. Stein, J. (2017). PV system performance modeling. Technical report. Sandia National Lab.(SNL-NM), Albuquerque, NM (USA).
17. Stutzer, M. J. (1980). Chaotic dynamics and bifurcation in a macro model. *Journal of Economic Dynamics and Control, 2*, 353–376.
18. Tacha, O. I., et al. (2018). Determining the chaotic behavior in a fractional-order finance system with negative parameters. *Nonlinear Dynamics, 94.2*, 1303–1317.
19. Tanaka, G., et al. (2021). Reservoir computing with diverse timescales for prediction of multiscale dynamics. arXiv:2108.09446.
20. Upadhyay, R. K. (2000). Chaotic behaviour of population dynamic systems in ecology. *Mathematical and Computer Modelling, 32.9*, 1005–1015.
21. Van Truc, N., & Anh, D. T. (2018). Chaotic time series prediction using radial basis function networks. In *4th International Conference on Green Technology and Sustainable Development (GTSD)* (pp. 753–758).

Chapter 4
Neural Approaches for Time Series Forecasting

Abstract The problem of forecasting a time series with a neural network is well-defined when considering a single step-ahead prediction. The situation becomes more tangled in the prediction on a multiple-step horizon and consequently the task can be framed in different ways. For example, one can develop a single-step predictor to be used recursively along the forecasting horizon (recursive approach) or develop a multi-output model that directly forecasts the entire sequence of output values (multi-output approach). Additionally, the internal structure of each predictor may be constituted by a classical feed-forward (FF) or by a recurrent architecture, such as the long short-term memory (LSTM) nets. The latter are traditionally trained with the teacher forcing algorithm (LSTM-TF) to speed up the convergence of the optimization, or without it (LSTM-no-TF), in order to avoid the issue of exposure bias. Time series forecasting requires organizing the available data into input-output sequences for parameter training, hyperparameter tuning and performance testing. An additional developers' choice explored in the chapter is the definition of the similarity index (error metric) that the training procedure must optimize and the other performance indicators that may be used to examine how well the prediction replicates test data.

The problem of forecasting h steps ahead (leads) of a univariate time series $y(t)$ consists of identifying a predictor that takes as an input the m last samples (lags), $y(t - m + 1), \ldots, y(t - 1), y(t)$, and returns as an output the predictions $\hat{y}(t + 1)$, $\hat{y}(t + 2), \ldots, \hat{y}(t + h)$ of the values $y(t + 1), y(t + 2), \ldots, y(t + h)$. In the machine learning terminology, we say that the time series forecasting has to be framed as a supervised learning problem [2]. The procedure to solve such a problem consists of reorganizing the data into a matrix of inputs (N rows and m columns) and a corresponding target output matrix (N rows and h columns) as shown in Fig. 4.1.

The number m of lags defines how many autoregressive terms the model takes as input. In order to make the learning problem feasible, this number must be large enough to establish a relationship between the input and the one-step ahead prediction $\hat{y}(t + 1)$. In nonlinear time series analysis, a suitable m is said to allow the embedding of the dataset and the minimum m is called the dataset embedding dimension. When the time series is generated by a known (autonomous, time-invariant) chaotic system,

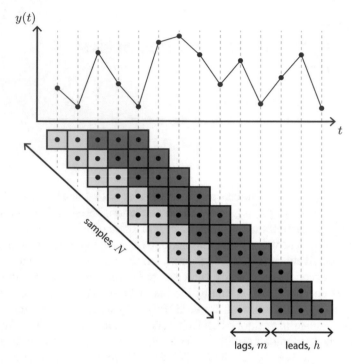

Fig. 4.1 The time series prediction is framed as a supervised learning task. For illustrative purposes, we represent N pairs of inputs ($m = 2$ steps, in yellow) and outputs ($h = 3$ steps, in blue). Note that the number of input-output pairs, N, is equal to the number of points in the time series minus $(m + h - 1)$

the embedding theory says that the smallest integer larger than twice the fractal dimension of the system's attractor generally allows the embedding [23]. Moreover, if the system is formulated as a m-dimensional nonlinear regression of the output variable $y(t)$, i.e., $y(t + 1)$ is expressed as a nonlinear function of the lags $y(t - m + 1), \ldots, y(t - 1), y(t)$, then m is the embedding dimension. This is the case of the chaotic oscillators on which we test our predictors. For real-world datasets, the embedding dimension can be numerically estimated [8]. Too large values of m can lead to a worse performance of the predictor, as further discussed in Subsect. 5.1.2.

4.1   Neural Approaches for Time Series Prediction

We focus on neural architectures formed by feed-forward and recurrent neurons. First, it is important to underline that, in principle, both FF nets and RNNs are able to solve the forecasting problem defined above. The difference is that FF predictors see the

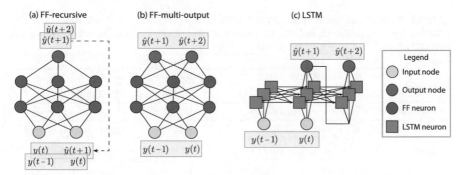

Fig. 4.2 Neural architectures (shown for $m = 2$ and $h = 2$). For illustrative purposes, we represent 2 hidden layers of 3 neurons each for the FF architectures and a single layer with 3 nodes for the LSTM

problem as a static one (they try to reproduce the relationship between the input and output sets), while RNNs look at the same problem from a dynamical perspective. In both cases, it is necessary to reorganize the time series, $y(1), y(2), \ldots, y(t), \ldots$, in N pairs of input $\left[y(t - m + 1), \ldots, y(t - 1), y(t)\right]$, and output $\left[y(t + 1), y(t + 2), \ldots, y(t + h)\right]$ vectors, called input-output samples in the machine learning terminology. The predictor is then a function from the m-dimensional input space to the h-dimensional output space.

The three neural architectures we focus on, FF-recursive and FF-multi-output used as benchmark and LSTM, are reported in Fig. 4.2 and will be described in detail in the following subsections.

The feed-forward models have been implemented using Keras [3] with Tensorflow as backend and the LSTM network using PyTorch [15].

4.1.1 FF-Recursive Predictor

The most common and natural forecasting approach consists of identifying the best single-step ahead predictor and then using it in a recursive way, feeding the previous step prediction back into the input vector of the following step (see Fig. 4.2a and Table 4.1). This approach is called recursive [6, 20] or iterative [2, 10, 14, 18]. Note that despite the dynamic nature of the time series, the identification of an FF-recursive predictor is a static task. It basically requires mimicking the mapping from m inputs to a single output.

The main advantage of this approach is that once the one-step predictor has been trained, it can be used recursively to forecast an arbitrarily long sequence. This is, however, its main drawback: it is not optimized for the task we are dealing with, which is a multi-step prediction.

Table 4.1 Training and inference phases of the four neural predictors for $m = 2$ and $h = 2$. $y(t)$ is the variable to be predicted, and $s(t)$ the internal state of the recurrent architecture. $s(t - m)$ is the initial internal state vector, whose elements are set to 0

Predictor	Training phase	Inference phase
FF-recursive	$\hat{y}(t+1) = f_{\text{FF-rec}}\big(y(t), y(t-1)\big)$	$\hat{y}(t+1) = f_{\text{FF-rec}}\big(y(t), y(t-1)\big)$
	$\hat{y}(t+2) = f_{\text{FF-rec}}\big(y(t+1), y(t)\big)$	$\hat{y}(t+2) = f_{\text{FF-rec}}\big(\hat{y}(t+1), y(t)\big)$
FF-multi-output	$\big[\hat{y}(t+2), \hat{y}(t+1)\big] = f_{\text{FF-mo}}\big(y(t), y(t-1)\big)$	$\big[\hat{y}(t+2), \hat{y}(t+1)\big] = f_{\text{FF-mo}}\big(y(t), y(t-1)\big)$
LSTM-TF	$\big[\hat{y}(t), s(t-1)\big] = f_{\text{LSTM-TF}}\big(y(t-1), s(t-2)\big)$	$\big[\hat{y}(t), s(t-1)\big] = f_{\text{LSTM-TF}}\big(y(t-1), s(t-2)\big)$
	$\big[\hat{y}(t+1), s(t)\big] = f_{\text{LSTM-TF}}\big(y(t), s(t-1)\big)$	$\big[\hat{y}(t+1), s(t)\big] = f_{\text{LSTM-TF}}\big(y(t), s(t-1)\big)$
	$\big[\hat{y}(t+2), s(t+1)\big] = f_{\text{LSTM-TF}}\big(y(t+1), s(t)\big)$	$\big[\hat{y}(t+2), s(t+1)\big] = f_{\text{LSTM-TF}}\big(\hat{y}(t+1), s(t)\big)$
LSTM-no-TF	$\big[\hat{y}(t), s(t-1)\big] = f_{\text{LSTM-no-TF}}\big(y(t-1), s(t-2)\big)$	$\big[\hat{y}(t), s(t-1)\big] = f_{\text{LSTM-no-TF}}\big(y(t-1), s(t-2)\big)$
	$\big[\hat{y}(t+1), s(t)\big] = f_{\text{LSTM-no-TF}}\big(y(t), s(t-1)\big)$	$\big[\hat{y}(t+1), s(t)\big] = f_{\text{LSTM-no-TF}}\big(y(t), s(t-1)\big)$
	$\big[\hat{y}(t+2), s(t+1)\big] = f_{\text{LSTM-no-TF}}\big(\hat{y}(t+1), s(t)\big)$	$\big[\hat{y}(t+2), s(t+1)\big] = f_{\text{LSTM-no-TF}}\big(\hat{y}(t+1), s(t)\big)$

4.1.2   FF-Multi-Output Predictor

In order to optimize the prediction on the entire sequence, it is necessary to define a model with multiple outputs. It can be done adopting a feed-forward and fully-connected architecture (Fig. 4.2b and Table 4.1). The structure is the same already used for the single-step predictor with the only difference in the output layer, whose number of nodes increases from 1 to h (multi-output [6, 20, 21] or direct [10] approach). Each neuron in the output layer focuses on the prediction of the considered variable at a different time step. The main issue with this architecture is that it does not explicitly take into account the fact that the outputs are sequential (i.e., the same variable at different time steps). In fact, the model would act in the same way if the outputs were to predict different system's variables at the same time step.

4.1.3   LSTM Predictor

The third ANN we consider is rather similar to the sequence-to-sequence architecture, which obtained outstanding performances in natural language processing in the last decade [22]. The idea, schematically represented in Fig. 4.2c, is to set up a structure capable of explicitly taking into account the sequentiality of the time series. To do so, the nodes in the hidden layers are not traditional feed-forward neurons but recurrent ones, in particular LSTM cells (Fig. 4.3). Each LSTM cell has two internal states (named "hidden" and "cell" state, respectively) and three gates (input, output and forget gates) [5, 7]. Intuitively, the cell state is responsible for keeping track of the relevant information provided by past inputs; the hidden state synthesizes the

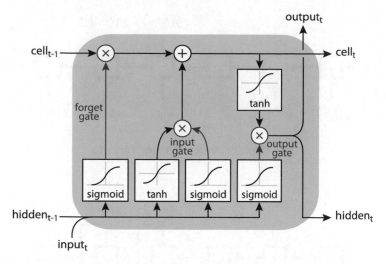

Fig. 4.3 Schematic diagram of an LSTM cell. The input, forget and output gates are represented in red

information provided by the current input, the cell state and the previous hidden state. Note that the hidden state is also the LSTM's output and corresponds to the state of a standard recurrent neuron. The gates regulate the flow of information in the neuron. Each gate's value is the output of a sigmoid in the range from 0 (closed gate) to 1 (open gate). The forget gate enables the LSTM neuron to reset the cell state at appropriate times (when the combination of input and hidden state into the sigmoid is low). Similarly, the input gate decides the extent to which the candidate cell state update (generated by a hyperbolic tangent) affects the cell state, protecting it from irrelevant contributions. The output gate modulates the candidate hidden state (generated by the other hyperbolic tangent) and avoids the negative effect of currently irrelevant memory contents. During training, this gated structure prevents vanishing (or exploding) gradients that generally affect convergence in traditional RNNs.

The distinctive feature of RNNs is that the weights of the neurons that process different time steps are shared (parameters sharing). For this reason, the number of parameters does not change if one considers more (or less) autoregressive terms in the past (lags m), or a different number of predicted steps ahead (leads h). In principle, this should be an advantage when m and/or h are large. At the same time, the weights shared across different time steps make the optimization problem more complex, which hence requires higher computational effort. This architecture potentially overcomes the limitations of the FF-recursive and multi-output predictors, because it is trained to reproduce the entire sequence of output variables and explicitly takes into account the fact that the h outputs are sequential values of the same variable.

The RNNs are usually trained using a technique known as "teacher forcing" (TF) [25]. It consists of using the target data as the input for each time step ahead, rather than the output predicted by the network at the previous step, as shown in

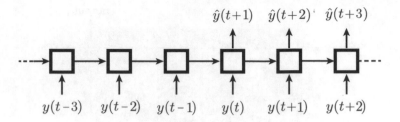

(a) Training with teacher forcing

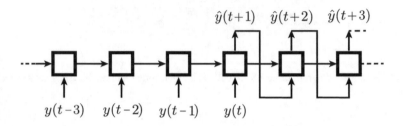

(b) Training without teacher forcing

Fig. 4.4 LSTM nets training procedure with (**a**) and without (**b**) teacher forcing

Fig. 4.4a and in Table 4.1. This technique proved to be efficient in various tasks related to natural language processing, granting faster convergence and avoiding the accumulation of errors in the initial phase of training [1]. For these reasons, all the high-level application programming interfaces (APIs) for deep learning adopt the TF by default [12]. To implement a training without TF (see Fig. 4.4b and Table 4.1), it is necessary to code a non-standard implementation directly in low-level APIs, such as TensorFlow or PyTorch.

In inference mode, when forecasting multiple time steps, the unknown previous value is replaced by the model prediction. TF therefore introduces a discrepancy between training and inference (exposure bias), yielding errors that accumulate along the sequence of predictions [1, 17]. Training with TF does not allow the network to correct its own mistakes because, during the training phase, the prediction at a certain time step does not affect future predictions as it does during inference. In forecasting tasks, this leads to a situation that is somewhat similar to use a one-step, recursive predictor.

We thus propose to train the LSTM architecture without TF (LSTM-no-TF). Coupling these two elements, recurrent neurons and no-TF, solves at the same time the drawbacks of the FF-recursive and multi-output predictors and of LSTM-TF. Indeed, this architecture is trained to predict the entire sequence of output variables and it explicitly takes into account the temporal connection between these outputs.

Thanks to the no-TF method, the network behaviors in training and inference modes coincide, so that the network can correct the prediction errors that can propagate during training [20].

In order to have a complete overview of the difference between training a recurrent architecture with and without TF, we analyze the forward and backward pass in a simplified neural architecture.

Consider a low-dimensional case with $m = 1$ and $h = 2$ and a single-layer structure. The network has three sets of parameters: w_i indicates the input to hidden state weights, w_s the connections between the internal hidden states, and w_o the state to output parameters.

The overall error to be minimized (E) is computed summing up the contributions at different steps:

$$E = E(t) + E(t+1) = \mathcal{L}\big(y(t+1), \hat{y}(t+1)\big) + \mathcal{L}\big(y(t+2), \hat{y}(t+2)\big) \quad (4.1)$$

where $\mathcal{L}\big(y(t), \hat{y}(t)\big)$ indicates a generic loss function measuring the distance between target and predicted outputs.

The training procedure requires the gradient of E to be computed with respect to the network parameters at the different time steps. The gradients corresponding to the same parameter at different steps are summed up together since we need a single gradient for each parameter in order to update its value at each iteration of the optimization algorithm.

For illustrative purposes, we show here the computation of $\frac{\partial E}{\partial w_i(t)}$ with TF and no-TF. The former is $\left[\frac{\partial E}{\partial w_i(t)}\right]^{ss}$, where the superscript ss indicates that the derivative is computed following the state-to-state loop only, as it happens with TF. The latter is $\left[\frac{\partial E}{\partial w_i(t)}\right]^{oi+ss}$, because no-TF considers the output-to-input as well as the state-to-state loop.

First, we consider the training with TF. Based on the schematic representation in Fig. 4.5, and making use of the chain rule, the equation (4.1) can be rewritten as:

$$\left[\frac{\partial E}{\partial w_i(t)}\right]^{ss} = \frac{\partial E(t)}{\partial w_i(t)} + \left[\frac{\partial E(t+1)}{\partial w_i(t)}\right]^{ss}$$
$$= \frac{\partial E(t)}{\partial s(t)} \cdot \frac{\partial s(t)}{\partial w_i(t)} + \frac{\partial E(t+1)}{\partial s(t+1)} \cdot \left[\frac{\partial s(t+1)}{\partial s(t)}\right]^{ss} \cdot \frac{\partial s(t)}{\partial w_i(t)}. \quad (4.2)$$

The gradient depends on two terms. The first takes into account the error at time t, while the second that at time $t + 1$.

$\left[\frac{\partial s(t+1)}{\partial s(t)}\right]^{ss}$ is the critical factor of the chain, since it requires the nonlinear activation function $\sigma(\cdot)$ to be derived (e.g., a sigmoid or a hyperbolic tangent):

$$s(t+1) = \sigma\Big(w_i(t+1)y(t+1) + w_s(t+1)s(t)\Big). \quad (4.3)$$

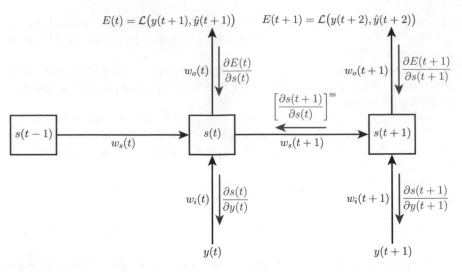

Fig. 4.5 Forward (black) and backward pass (red) in a recurrent architecture trained with teacher forcing

This expression contains the key of TF, since the new internal state $s(t + 1)$ depends on the previous state $s(t)$ and on the actual value $y(t + 1)$. The latter does not depend on the network weights, and thus it interrupts the derivatives chain. Making few basic computations, we can expand the derivative of $s(t + 1)$ with respect to $s(t)$ as follows:

$$
\begin{aligned}
\left[\frac{\partial s(t + 1)}{\partial s(t)}\right]^{\text{ss}} &= \frac{\partial}{\partial s(t)}\sigma\Big(w_i(t + 1)y(t + 1) + w_s(t + 1)s(t)\Big) \\
&= \sigma'\Big(w_i(t + 1)y(t + 1) + w_s(t + 1)s(t)\Big) \cdot \\
&\quad \frac{\partial}{\partial s(t)}\big(w_i(t + 1)y(t + 1) + w_s(t + 1)s(t)\big) \\
&= \sigma'\Big(w_i(t + 1)y(t + 1) + w_s(t + 1)s(t)\Big) \cdot w_s(t + 1),
\end{aligned}
\tag{4.4}
$$

being $\sigma'(\cdot)$ the derivative of the function $\sigma(\cdot)$ with respect to its argument.

Considering a training with no-TF, the equation (4.3) changes because the actual value $y(t + 1)$ is replaced with the prediction $\hat{y}(t + 1)$, which is dependent on the weights at the previous time steps:

$$
\begin{aligned}
s(t + 1) &= \sigma\Big(w_i(t + 1)\hat{y}(t + 1) + w_s(t + 1)s(t)\Big) \\
&= \sigma\Big(w_i(t + 1)w_o(t)s(t) + w_s(t + 1)s(t)\Big).
\end{aligned}
\tag{4.5}
$$

As represented in Fig. 4.6, $s(t + 1)$ has a two-fold dependence on $s(t)$: due to the state-to-state and the output-to-input loop.

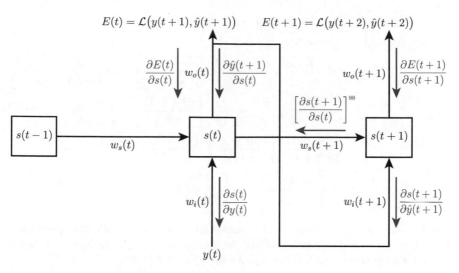

Fig. 4.6 Forward (black) and backward pass (red) in a recurrent architecture trained without teacher forcing

The derivative can be computed as:

$$
\begin{aligned}
\left[\frac{\partial s(t+1)}{\partial s(t)}\right]^{\text{oi+ss}} &= \frac{\partial}{\partial s(t)}\sigma\Big(w_i(t+1)w_o(t)s(t)+w_s(t+1)s(t)\Big) \\
&= \sigma'\Big(w_i(t+1)w_o(t)s(t)+w_s(t+1)s(t)\Big)\cdot \\
&\quad \frac{\partial}{\partial s(t)}\big(w_i(t+1)w_o(t)s(t)+w_s(t+1)s(t)\big) \\
&= \sigma'\Big(w_i(t+1)w_o(t)s(t)+w_s(t+1)s(t)\Big)\cdot \\
&\quad \big(w_i(t+1)w_o(t)+w_s(t+1)\big) \\
&= \frac{\partial}{\partial s(t)}\big(w_i(t+1)w_o(t)s(t)+w_s(t+1)s(t)\big)w_i(t+1)w_o(t)+ \\
&\quad \frac{\partial}{\partial s(t)}\big(w_i(t+1)w_o(t)s(t)+w_s(t+1)s(t)\big)w_s(t+1) \\
&= \left[\frac{\partial s(t+1)}{\partial s(t)}\right]^{\text{oi}} + \left[\frac{\partial s(t+1)}{\partial s(t)}\right]^{\text{ss}}.
\end{aligned}
\tag{4.6}
$$

Equations (4.4) and (4.6) report the expressions of the state-to-state gradient computed with and without TF.

Analogously, it is possible to compute the derivative of the loss function with respect to the considered weight with TF and no-TF.

$$
\begin{aligned}
\left[\frac{\partial E}{\partial w_i(t)}\right]^{oi+ss} &= \frac{\partial E(t)}{\partial w_i(t)} + \left[\frac{\partial E(t+1)}{\partial w_i(t)}\right]^{oi+ss} \\
&= \frac{\partial E(t)}{\partial w_i(t)} + \frac{\partial E(t+1)}{\partial s(t+1)} \cdot \left[\frac{\partial s(t+1)}{\partial s(t)}\right]^{oi+ss} \cdot \frac{\partial s(t)}{\partial w_i(t)} \\
&= \frac{\partial E(t)}{\partial w_i(t)} + \frac{\partial E(t+1)}{\partial s(t+1)} \cdot \left[\frac{\partial s(t+1)}{\partial s(t)}\right]^{oi} \cdot \frac{\partial s(t)}{\partial w_i(t)} + \\
&\quad \frac{\partial E(t+1)}{\partial s(t+1)} \cdot \left[\frac{\partial s(t+1)}{\partial s(t)}\right]^{ss} \cdot \frac{\partial s(t)}{\partial w_i(t)} \\
&= \left[\frac{\partial E(t+1)}{\partial w_i(t)}\right]^{oi} + \left[\frac{\partial E}{\partial w_i(t)}\right]^{ss}.
\end{aligned} \tag{4.7}
$$

The gradients computed with no-TF is composed by two terms: the first specifically takes into account the presence of the recurence from output to input, while the second is exactly the gradient computed with TF (see Eq. 4.2).

The specific computation of $\frac{\partial E}{\partial w_i(t)}$ can be easily generalized to the other parameters. Note that the gradients relative to time step $t+1$ (i.e., w_{t+1}^i, w_{t+1}^s and w_{t+1}^o), which in this case is the last step, are computed in the same way in TF and no-TF.

As it is easy to understand from the simple example presented above, the analytical derivation of the gradients is not trivial, especially when considering complex neurons (such as the LSTM cells) and a high number of unfolding steps. For this reason, the modern deep learning libraries such as PyTorch, TensorFlow and Keras do not require the user to derive the gradients explicitly. In fact, one has to define only the forward pass (i.e., the topology of the computational graph to calculate the outputs starting from the inputs) and to specify the loss function. Based on these elements, the software automatically estimates the gradients required for the optimization process.

4.2   Performance Metrics

The predictions obtained with the proposed ANNs have to be compared with the target values that are computed by simulating the chaotic system. The comparison is done using appropriate metrics. A traditional metric for regression tasks is the mean squared error (MSE, hereafter). Focusing on the forecasting of the ith step ahead, we can consider N target samples $\mathbf{y} = [y_1, y_2, \ldots, y_N]$, not necessarily in temporal order, and the corresponding i-step ahead predictions $\hat{\mathbf{y}}^{(i)} = [\hat{y}_1^{(i)}, \hat{y}_2^{(i)}, \ldots, \hat{y}_N^{(i)}]$, i.e., $y_k = y(t_k + i)$ for some t_k in the dataset and $\hat{y}_k^{(i)}$ is the predictor output $\hat{y}(t_k + i)$ computed for the input $[y(t_k), y(t_k - 1), \ldots, y(t_k - m + 1)]$.

The MSE is then defined as:

$$
\text{MSE}(\mathbf{y}, \hat{\mathbf{y}}^{(i)}) = \frac{1}{N} \sum_{k=1}^{N} \left(y_k - \hat{y}_k^{(i)}\right)^2. \tag{4.8}
$$

This is the metric typically used as a loss function to be minimized during the network training. The main disadvantage of the MSE is that its numerical values do not provide a good insight on the quality of the prediction, because they are not normalized with respect to the variability of the data. For this reason, to assess the prediction quality, it is preferable to use a relative metric, such as the R^2 score, sometimes known as the Nash–Sutcliffe model efficiency coefficient [11, 13] (not to be confused with the square of Pearson correlation coefficient). It is defined by the following expression:

$$R^2\left(\mathbf{y}, \hat{\mathbf{y}}^{(i)}\right) = 1 - \frac{\mathrm{MSE}\left(\mathbf{y}, \hat{\mathbf{y}}^{(i)}\right)}{\mathrm{MSE}(\mathbf{y}, \bar{y})} = 1 - \frac{\sum_{k=1}^{N}\left(y_k - \hat{y}_k^{(i)}\right)^2}{\sum_{k=1}^{N}\left(y_k - \bar{y}\right)^2}, \tag{4.9}$$

where $\bar{y}$ is the mean of the target data. The R^2 score measures the predictive power of a given model with respect to the predictive power of the trivial model which always forecasts the mean value of the observed data (in this case, $R^2 = 0$). This metric varies in the range $(-\infty, 1]$, with the upper bound corresponding to a perfect forecasting. The R^2 score is widely used because it can be seen as a normalized version of the MSE, $\mathrm{MSE}(\mathbf{y}, \bar{y})$ being the data variance.

The MSE and R^2 score, defined here for the prediction of the ith step, can be averaged over the whole forecasting horizon of h steps, i.e.,

$$\langle \mathrm{MSE} \rangle = \frac{1}{h}\sum_{i=1}^{h}\mathrm{MSE}\left(\mathbf{y}, \hat{\mathbf{y}}^{(i)}\right)$$

$$\langle R^2 \rangle = \frac{1}{h}\sum_{i=1}^{h}R^2\left(\mathbf{y}, \hat{\mathbf{y}}^{(i)}\right), \tag{4.10}$$

Specifically for the prediction of chaotic systems, we introduce a third metric based on the system's LLE. It is defined as the number of Lyapunov times ahead for which the prediction error does not exceed a given threshold, where the Lyapunov time (LT, see Subsect. 2.3.1) is the inverse of the LLE. This metric normalizes the prediction power of the neural model to the average chaoticity of the data generator and allows a fair comparison of the predictor's performance across different chaotic systems [24].

In chaotic systems, even an ideal predictor would diverge sooner or later from the original output values because the local divergence that typifies chaos (the LLE is the average exponential rate of local divergence) amplifies any small numerical error. In this case, if the system is well approximated by the predictor, the R^2 score converges to -1 [4, 19]. To demonstrate this, we assume that the same statistical properties hold for the target output of the system, $\mathbf{y}$, and its prediction, $\hat{\mathbf{y}}^{(h)}$:

$$E[\mathbf{y}] = E[\hat{\mathbf{y}}^{(h)}] = \bar{y} \tag{4.11}$$

$$\text{Var}[\mathbf{y}] = \text{Var}[\hat{\mathbf{y}}^{(h)}] \tag{4.12}$$

$$E[\mathbf{y}^2] = E[(\hat{\mathbf{y}}^{(h)})^2], \tag{4.13}$$

where $E[\cdot]$ denotes the mean operator and $\text{Var}[\cdot]$ the variance. After divergence, we can imagine the target and the prediction as two uncorrelated trajectories within the oscillator's attractor. In other words, the two variables become independent. We can thus make use of the following well-known statistical property:

$$E[\mathbf{y} \cdot \hat{\mathbf{y}}^{(h)}] = E[\mathbf{y}] \cdot E[\hat{\mathbf{y}}^{(h)}]. \tag{4.14}$$

Based on (4.9), the R^2 score is given by:

$$R^2\left(\mathbf{y}, \hat{\mathbf{y}}^{(h)}\right) = 1 - \frac{E[(\mathbf{y} - \hat{\mathbf{y}}^{(h)})^2]}{E[(\mathbf{y} - \bar{y})^2]}. \tag{4.15}$$

The variance can be rephrased as:

$$\begin{aligned}
E[(\mathbf{y} - \bar{y})^2] &= E[\mathbf{y}^2 - 2\mathbf{y}\bar{y} + \bar{y}^2] \\
&= E[\mathbf{y}^2] - 2E[\mathbf{y}\bar{y}] + E[\bar{y}^2] \\
&= E[\mathbf{y}^2] - 2\bar{y}^2 + \bar{y}^2 \\
&= E[\mathbf{y}^2] - \bar{y}^2.
\end{aligned} \tag{4.16}$$

The mean squared error, $\text{MSE}\left(\mathbf{y}, \hat{\mathbf{y}}^{(h)}\right)$, in turn can be formulated as:

$$\begin{aligned}
E[(\mathbf{y} - \hat{\mathbf{y}}^{(h)})^2] &= E[\mathbf{y}^2 - 2\mathbf{y}\hat{\mathbf{y}}^{(h)} + (\hat{\mathbf{y}}^{(h)})^2] \\
&= E[\mathbf{y}^2] - E[2\mathbf{y}\hat{\mathbf{y}}^{(h)}] + E[(\hat{\mathbf{y}}^{(h)})^2] \\
&= E[\mathbf{y}^2] - 2E[\mathbf{y}]E[\hat{\mathbf{y}}^{(h)}] + E[(\hat{\mathbf{y}}^{(h)})^2] \\
&= 2E[\mathbf{y}^2] - 2E[\bar{y}^2] \\
&= 2(E[\mathbf{y}^2] - \bar{y}^2).
\end{aligned} \tag{4.17}$$

Having shown that for independent datasets with identical statistical properties (same mean, variance and second raw moment) the MSE of two independent trajectories averages out to the double of the variance of the process, the R^2 score converges as:

$$R^2\left(\mathbf{y}, \hat{\mathbf{y}}^{(h)}\right) = 1 - \frac{2(E[\mathbf{y}^2] - \bar{y}^2)}{E[\mathbf{y}^2] - \bar{y}^2} = 1 - 2 = -1. \tag{4.18}$$

The fact that the R^2 score approaches -1 for large h is hence a quality check for the predictor. It means that the predictor, used recursively as a dynamical system itself, is able to generate an attractor with the same characteristics (i.e., the same statistical properties) of the data generator. Some authors say that the model has the capability of replicating the chaotic attractor's long-term "climate" [16]. This property is not interesting while considering a short-term forecasting horizon. It becomes relevant if one wants to use the model identified for some other related but different tasks, such as the generation of synthetic series, or the computation of the Lyapunov exponents (see Sect. 6.2).

4.3 Training Procedure

Each artificial system is used to generate 50,000 data points, which have been split into training (70% of the samples), validation (15%), and test set (15%). Considering the two real-world time series, solar irradiance and ozone concentration, it is not possible to generate an arbitrary number of samples. Depending on the amount of data available, we keep one or two whole years for validation (the same goes for testing) in order to preserve the annual periodicity of the natural processes.

The training set, as the name suggests, is used to train the parameters (i.e., weights and bias) of the neural network. The input-output pairs of the validation set are not directly processed by the optimization algorithm, but are used to tune the hyperparameters of the neural network, which define the network structure and the learning algorithm configuration (see Table 4.2). The test set does not have a role in the model identification; it is just used to evaluate the model's performance in inference mode fairly.

In order to limit the number of hyperparameters to be tuned on the one hand, and to provide a fair comparison between different architectures on the other, we fixed the number of hidden layers (3 layers) and the number of neurons per layer (10 neurons). Other hyperparameters (see Table 4.2) have been tuned through a grid search optimization. We tested a batch size of 256, 512 and 1024; learning rates equal to 0.1, 0.01 and 0.001; decay factor of 0.001 and 0. The number of epochs was fixed to 5000, a value which ensures the convergence of each training. Both training and validation losses are monitored during the learning process, so that

Table 4.2 Description of the neural network's hyperparameter (so-called to avoid confusion with the network parameters, weights and biases, which define the input-output mapping)

Hyperparameter	Description
Batch size	Number of training input-output samples used to compute the loss function gradient in one iteration of the optimization
Learning rate	Step along the gradient by which the network's parameters are updated at each iteration of the optimization (also called step size)
L-rate decay	Exponential rate of decay of the learning rate during training
Num. of epochs	Number of times the learning algorithm iterates through the entire training dataset, with sample reshuffling at each iteration

the parameterization selected for a specific combination of the hyperparameters is the one which has the best performance on the validation dataset (instead of the one obtained in the last epoch). For each combination of the hyperparameters, the optimization is repeated twice, varying the random weight initialization in order to avoid dependence on particularly unlucky initial values. The training has been performed using Adam (adaptive moment estimation) optimizer, [9] a state-of-the-art algorithm for deep learning derived from the standard stochastic gradient descent algorithm that is extensively used in deep learning applications. As suggested by its name, Adam adaptively estimates the first- and second-order moments of the gradients using running averages with decay rates equal to $\beta_1 = 0.9$ and $\beta_2 = 0.999$ for the first and second moments, respectively. The estimates of the two moments are then used to evolve the learning rate during the training procedure.

References

1. Bengio, S., et al. (2015). Scheduled sampling for sequence prediction with recurrent neural networks. *Proceedings of the 29th Conference on Neural Information Processing Systems*, 28, 1171–1179.
2. Bontempi, G., Ben Taieb, S., & Le Borgne, Y.-A. (2012). Machine learning strategies for time series forecasting. In *European business intelligence summer school* (pp. 62–77). Springer.
3. Chollet, F., et al. (2018). Keras: The python deep learning library. *Astro- physics Source Code Library*.
4. Dercole, F., Sangiorgio, M., & Schmirander, Y. (2020). An empirical assessment of the universality of ANNs to predict oscillatory time series. *IFAC-PapersOnLine, 53.2*, 1255–1260.
5. Goodfellow, I., Bengio, Y., & Courville, A. (2015). Deep learning. MIT Press.
6. Guariso, G., Nunnari, G., & Sangiorgio, M. (2020). Multi-step solar irradiance forecasting and domain adaptation of deep neural networks. *Energies, 13.15*, 3987.
7. Hochreiter, S., & Schmidhuber, J. (1997). Long short-term memory. *Neural computation, 9.8*, 1735–1780.
8. Kennel, M. B., Brown, R., & Abarbanel, H. D. (1992). Determining embedding dimension for phase-space reconstruction using a geometrical construction. *Physical Review A, 45.6*, 3403.
9. Kingma, D. P., & Ba, J. (2014). Adam: A method for stochastic optimization. arXiv:1412.6980.
10. Makridakis, S., Spiliotis, E., & Assimakopoulos, V. (2018). Statistical and Machine Learning forecasting methods: Concerns and ways forward. *PloS one, 13.3*, e0194889.
11. McCuen, R. H., Knight, Z., & Cutter, G. (2006). Evaluation of the Nash-Sutcliffe efficiency index. *Journal of Hydrologic Engineering, 11.6* , 597–602.
12. Mihaylova, T., & Martins, A. F. T. (2019). Scheduled Sampling for Transformers. arXiv:1906.07651.
13. Nash, J. E., & Sutcliffe, J. V. (1970). River flow forecasting through conceptual models part I-A discussion of principles. *Journal of Hydrology, 10.3*, pp. 282–290.
14. Pan, S., & Duraisamy, K. (2018). Long-time predictive modeling of nonlinear dynamical systems using neural networks. *Complexity*.
15. Paszke, A., et al. (2017). Automatic differentiation in PyTorch. In *Proceedings of the Thirty-fifth Conference on Neural Information Processing Systems*.
16. Pathak, J., et al. (2017). Using machine learning to replicate chaotic attractors and calculate Lyapunov exponents from data. *Chaos: An Interdisciplinary Journal of Nonlinear Science, 27.12*, 121102.
17. Ranzato, M., et al. (2015). Sequence level training with recurrent neural networks. arXiv:1511.06732.

18. Rasp, S., et al. (2020). WeatherBench: A benchmark dataset for data-driven weather forecasting. *Journal of Advances in Modeling Earth Systems, 12.1*.
19. Sangiorgio, M. (2021). Deep learning in multi-step forecasting of chaotic dynamics. Ph.D. thesis. Department of Electronics, Information and Bioengineering, Politecnico di Milano.
20. Sangiorgio, M., & Dercole, F. (2020). Robustness of LSTM neural networks for multi-step forecasting of chaotic time series. *Chaos, Solitons and Fractals, 139*, 110045.
21. Sangiorgio, M., Dercole, F., & Guariso, G. (2021). Forecasting of noisy chaotic systems with deep neural networks. *Chaos, Solitons & Fractals*, 153, 111570.
22. Sutskever, I., Vinyals, O., & Le, Q.V. (2014). Sequence to sequence learning with neural networks. *Proceedings of the 28th Conference on Neural Information Processing Systems*, 27, 3104–3112.
23. Takens, F. (1981). Detecting strange attractors in turbulence. *Dynamical systems and turbulence, Warwick 1980* (pp. 366–381). Springer.
24. Wang, R., Kalnay, E., & Balachandran, B. (2019). Neural machine-based forecasting of chaotic dynamics. *Nonlinear Dynamics, 98.4*, 2903–2917.
25. Williams, R. J., & Zipser, D. (1989). A learning algorithm for continually running fully recurrent neural networks. *Neural Computation, 1.2*, 270–280.

Chapter 5
Neural Predictors' Accuracy

Abstract We examine the performance of different predictors in the deterministic environment and test their robustness against noise. In particular, we mimic the classical case of measurement noise by adding a random Gaussian signal of different intensity to the deterministic output of some archetypal chaotic systems. Then, we examine the critical case of structural noise, represented by the slow variation of the growth rate parameter of the logistic map. In both cases, the presence of noise rapidly degrades all the performance indicators, but, interestingly, the best deterministic predictor, i.e., LSTM trained without teacher forcing, remains the best also in the stochastic and non-stationary environments. Finally, we examine solar irradiance and ozone concentration time series, and again the same predictor turns out to be the best and can also be reliably applied to similar datasets in the same domain (domain adaptation).

In this chapter, we analyze the predictive power of the four considered neural predictors, FF-recursive, FF-multi-output, and LSTM with and without teacher forcing (LSTM-TF and LSTM-no-TF), on the time series presented in Chap. 3:

- the deterministic noise-free systems (logistic, Hénon and generalized Hénon maps);
- the noisy version of the same systems, considering three different levels of noise;
- a modified version of the logistic map, with a slow dynamic for the parameter;
- two real-world time series (i.e., solar irradiance and ozone concentration).

A detailed description of the experimental setting and some remarks on the training procedure are reported in Chap. 6.

5.1 Deterministic Systems

First, we train the FF-recursive predictor. As said in Subsect. 4.1.1, it is the most diffused way for predicting nonlinear time series and thus represents a benchmark. In addition, it allows us to define the length of the predictive horizon (i.e., the number h

Table 5.1 Features of the four considered chaotic systems

System	Lags (m)	Leads (h)	LT
Logistic	1	20	2.83
Hénon	2	17	2.33
3D generalized Hénon	3	27	3.62
10D generalized Hénon	10	109	15.87

of predicted time steps) for each chaotic system. Once the single-step predictor has been identified, we apply it recursively and compute the R^2 score for each time step ahead on the test dataset. We set the length h of the predicting horizon as the first step for which R^2 score becomes negative. The main features related to the forecasting task of the four chaotic systems are reported in Table 5.1.

Once h has been defined for each chaotic system, it is possible to train the multi-output and the LSTM predictors.

The R^2 score for each time step ahead during testing is reported in Fig. 5.1 for each predictor (lines on the same panel) and simulated system (same color line on different panels). The main result is that our four predictors rank in the same order in all cases. Leveraging on the wide range of complexity of the considered chaotic maps and of the corresponding forecasting tasks, we can assume this result to be a general property of the neural architectures taken into account. A second remarkable result is the rather uniform performance of each predictor across different systems, read in terms of the number of Lyapunov times (LTs) predicted with a sufficient R^2 score (quality threshold 0.7 in Table 5.2). Note that we intentionally designed our setup to challenge the predictors on tasks with different complexity. For this reason, we have used same-size networks and datasets to predict dynamics with different complexity. This is, however, an unfair setup for the comparison of the predictor performances on different tasks and, indeed, the accuracy of all predictors decays with the dimension and complexity of the forecasted system. Nonetheless, all neural architectures seem to show a universal predictive power, in line with the preliminary results in [3], where a fair comparison has been carried out for the FF-recursive predictor.

The FF-recursive predictor can mimic almost perfectly the behavior of the real systems for 5 LTs. After that, the performances rapidly degrade to the point that the R^2 score becomes negative (between 7 and 8 LTs). The FF-multi-output predictor definitely provides the poorest performances: the R^2 score drops after 2 LTs and then vanishes, showing that the predictions become close to the mean value of the training dataset. For the three low-dimensional systems, the maps to be approximated (the iterations of the system's map, see Fig. 3.2) become so complex for increasing lead time, that the network is not able to reproduce them (Fig. 5.1a–c). For the 10D generalized Hénon, the large horizon h (each LT corresponds to 15.87 time steps) is an additional issue. To perform similarly to the low-dimensional cases, the FF-multi-output predictor should reproduce more than 30 iterated maps independently, since the sequential nature of the outputs is not explicitly considered. This task is

Fig. 5.1 $R^2\left(\mathbf{y}, \hat{\mathbf{y}}^{(i)}\right)$ score obtained with the four predictors (see colored lines) on the logistic (**a**), Hénon (**b**), 3D generalized Hénon (**c**) and 10D generalized Hénon (**d**) maps. Performances computed on the noise-free test dataset

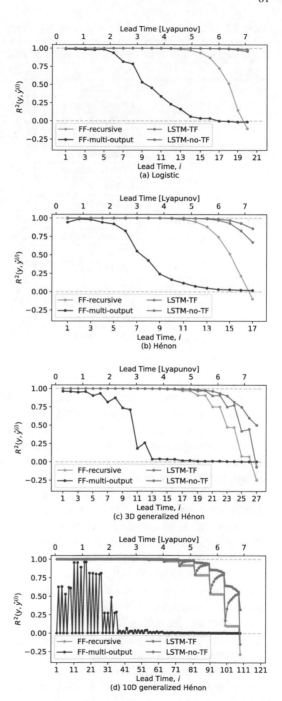

Table 5.2 Overall performance of the predictors over the four considered chaotic systems in the noise-free case. The third column reports the number of LTs after which the R^2 score goes below 0.7. In the fourth column we compute the average R^2 score on the h-step horizon

System	Predictor	# LT $R^2 > 0.7$	$\langle R^2 \rangle$
Logistic	FF-recursive	6.01	0.85
	FF-multi-output	2.83	0.46
	LSTM-TF	> 7.08	> 0.99
	LSTM-no-TF	> 7.08	> 0.99
Hénon	FF-recursive	6.00	0.83
	FF-multi-output	2.57	0.43
	LSTM-TF	6.86	0.96
	LSTM-no-TF	> 7.28	0.98
3D generalized Hénon	FF-recursive	6.07	0.82
	FF-multi-output	2.76	0.35
	LSTM-TF	6.62	0.89
	LSTM-no-TF	6.90	0.94
10D generalized Hénon	FF-recursive	5.67	0.84
	FF-multi-output	0.00	0.13
	LSTM-TF	5.67	0.90
	LSTM-no-TF	6.23	0.94

so complex that the optimization always gets stuck in local minima, where only a few non-necessarily consecutive steps are accurately predicted. FF-multi-output performances are thus acceptable only on short forecasting horizons (small h). The two LSTM predictors exhibit a trend similar to the FF-recursive one, always providing a higher precision, especially after 4–5 LTs. Training the LSTM without TF gives the best performance in all cases.

Note the behavior of the R^2 score in the two generalized Hénon maps (Fig. 5.1c–d), which is system-specific: the performance deteriorates step-wise while increasing the number of steps ahead. In fact, this step-wise behavior is a peculiar feature of the generalized Hénon map; more precisely, of recurrences in which the nonlinear map giving $y(t + 1)$ is applied only to a few last samples of the lag-window of m previous samples ($y(t - 8)$ and $y(t - 9)$ in the 10D generalized Hénon map). The first $m - 1$ steps ahead are forecasted using actual data (null error), required to initialize the predictor. After that, these $m - 1$ predictions (with a certain error), are used to forecast the values from m to $2 \cdot (m - 1)$ steps ahead, and so on.

The 10D map also highlights the relevance of the exposure bias problem (i.e., the discrepancy between training and inference modes) for LSTM networks with TF. The R^2 score of the LSTM-TF exhibits a structure organized in blocks of length $m - 1$, but the trend within each block is increasing. This counterintuitive behavior shows that the predictor particularly suffers at the beginning of each block because predictions of lower quality start to be used as an input, while this has not been

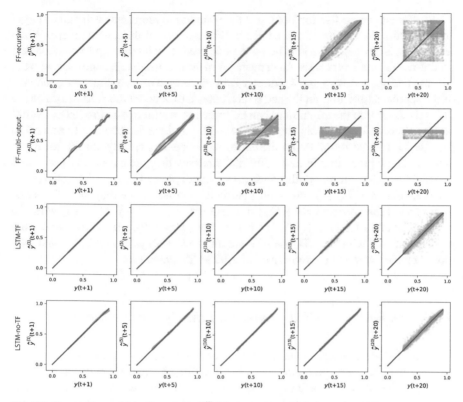

Fig. 5.2 Scatter plots of the points $(y(t), \hat{y}^{(i)}(t))$ for the logistic map, $i = 1, 5, 10, 15, 20$ (columns 1–5; different predictors in rows 1–4). Perfect predictions align on the diagonal

done in training. The propagation of the information through the hidden states then interestingly allows the predictor to recover until the next performance drop at the beginning of the next block.

Looking across the panels of Fig. 5.1 and at the results summarized in Table 5.2, it can be seen that the FF-recursive predictor is able to maintain rather similar performances on systems of different order. Conversely, the LSTM predictors seem to suffer more in the two generalized Hénon maps. Indeed, the forecasting task of a generalized Hénon map is more critical for a recurrent network than it is for an FF architecture. Because the prediction $\hat{y}(t + 1)$ does not depend on the input from $y(t), \ldots, y(t - m + 3)$, an FF net just has to nullify the corresponding weights. Conversely, an RNN has to perform a two-fold task: propagate the information through time over the m lags of the network (for the systems considered here, m equals the number of state variables) to remember past inputs and reproduce the nonlinear function. The results show that training the LSTM predictors without TF strongly mitigates this issue, supporting our hypothesis that this training method allows the proper information flow over subsequent time steps.

A different perspective for looking at the results is to represent each pair of target and prediction in a scatter plot (see Fig. 5.2 for the case of the logistic map). When forecasting is perfect, the points lie on the diagonal. It is interesting to note that the LSTM-no-TF predictor makes small mistakes in the first time step, compared with the TF case (compare rows 3 and 4, column 1, in Fig. 5.2). This is a direct consequence of the approach used to train the net, which takes into account the whole horizon, so that a small error in the initial steps can be somehow useful to obtain a better overall performance. In other words, the LSTM-no-TF predictor is specifically trained to provide the best performance on the entire sequence. Conversely, the FF-recursive and LSTM-TF predictors follow a different strategy: they are essentially optimized to predict one step ahead. For this reason, the scatter plots of the first time step (rows 1 and 3, column 1, Fig. 5.2) are almost perfectly aligned on the diagonal. The same trend can be seen even for $(y(t + 5), \hat{y}(t + 5))$ and $(y(t + 10), \hat{y}(t + 10))$, while the situation changes for $(y(t + 15), \hat{y}(t + 15))$ and $(y(t + 20), \hat{y}(t + 20))$, where the points generated with the LSTM-no-TF are much more aligned on the diagonal than those obtained with the FF-recursive and LSTM-TF predictors.

5.1.1   Performance Distribution over the System's Attractor

LLE and LT are average quantities which characterize the entire chaotic attractor. However, the same predictor could provide different performances in different regions of the attractor. For instance, starting a prediction near a saddle point typically affects the score of the prediction negatively [15]. Figure 5.3 shows the chaotic attractor of the 3D generalized Hénon system reconstructed via delay-coordinate embedding (which is equivalent to the state space in this case). The samples of the test set have been organized in triplets $[y(t), y(t-1), y(t-2)]$ which identify a point in the reconstructed state space. We then perform a forecasting of the following $h = 27$ steps ahead. The color assigned to each point depends on the performance obtained when the forecasting starts from the point itself, computed as:

$$R^2(y(t+1:t+h), \hat{y}^{(1:h)}(t+1:t+h)) =$$
$$R^2([y(t+1), \ldots, y(t+h)], [\hat{y}^{(1)}(t+1), \ldots, \hat{y}^{(h)}(t+h)]). \tag{5.1}$$

The attractor relative to the FF-recursive predictor (Fig. 5.3a) is mainly represented in warm colors (R^2 score close to 1). Points depicted with cold hues do not seem to have any particular distribution. The FF-multi-output network (Fig. 5.3b) confirms the low precision already observed in Fig. 5.1. The LSTM-TF (Fig. 5.3c) achieves better performances than the FF-recursive predictor, but still shows a similar random pattern. The best performance and uniformity is obtained with the LSTM-no-TF predictor (Fig. 5.3d).

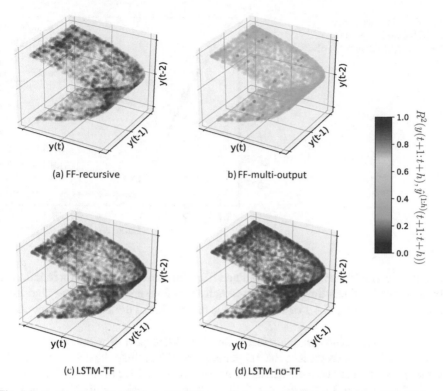

(a) FF-recursive

b) FF-multi-output

(c) LSTM-TF

(d) LSTM-no-TF

Fig. 5.3 Performance distribution of the four predictors over the chaotic attractor of the 3D generalized Hénon map. The color of each point indicates the 27-step forecasting performance when starting from that point

5.1.2 Sensitivity to the Embedding Dimension

Until now, we always assumed to have a mapping between m consecutive samples of the dataset (i.e., the dataset embedding dimension) and the following value. In the deterministic cases considered here, it is actually possible to compute m analytically from the systems' equations. However, this is not generally possible even when one knows the state-space representation of the system exactly. Moreover, the system's equations are typically unknown in real-world applications, so that m has to be identified by means of numerical techniques. One way, especially suitable for ANNs, consists of optimizing m as an additional hyperparameter [9, 10, 12, 13, 18]. Since hyperparameters' tuning is a time-consuming task, it is worthwhile to find a suitable set of values of m, in order to limit the computational effort. To this aim, Takens' theorem [17] is extremely useful, since it states that a chaotic system can be reconstructed from a time series provided that a sufficient (generically, only a little larger than twice the attractor fractal dimension) embedding dimension is used.

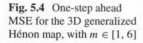

Fig. 5.4 One-step ahead MSE for the 3D generalized Hénon map, with $m \in [1, 6]$

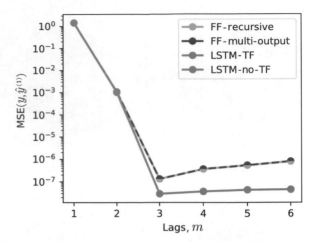

Alternatively, some insight can be obtained from the data auto-correlation function or by means of nonlinear time-series tools (such as the false nearest neighbor algorithm presented in Subsect. 2.4.1 or computing the dataset fractal dimension for increasing m [8]).

To investigate which predictor is more robust with respect to an uncertainty about the value of m, we reconsider the 3D generalized Hénon map. The attractor fractal dimension computed with the Kaplan-Yorke formula is 2.13. This means that, generically, 5 (the next integer larger than $2.13 \cdot 2$) lags at most are required to reconstruct the chaotic attractor via delay-coordinate embedding.

We first focus on the single-step prediction, comparing the performance of the FF and LSTM predictors (note that when $h = 1$, the recursive and multi-output FF predictors are equivalent, as well as the LSTM with or without TF). Figure 5.4 reports the one-step ahead MSE and shows that $m < 3$ is not sufficient: essentially there is no mapping between 2 consecutive output data and the next. As expected, $m = 3$ guarantees the best performance, in accordance with the conclusion analytically obtained from the equations of the model. When $m > 3$, the behavior of the LSTM predictors is nearly the same as the one obtained with only the necessary inputs. The extra available inputs have essentially no impact on the outputs, since they are filtered by the sequential LSTM architecture. This allows the predictor to correctly learn that they are not useful for improving the prediction. The FF structure appears to be less robust, because there is a static link between input and output. In principle, they should have the ability to avoid dependence on useless inputs, by setting to zero the weights of the connection between the redundant inputs and the first hidden layer. However, after the training, the value of these weights is close, but not exactly equal, to zero, so that the effect of useless inputs remains visible (note, however, the logarithmic scale in Fig. 5.4).

Figure 5.5 shows the average $\langle R^2 \rangle$ computed over 27 steps for different lags m (the same considered in Fig. 5.4). The LSTM-no-TF predictor again exhibits the best performance, both when the embedding dimension is underestimated ($m < 3$)

Fig. 5.5 R^2 score computed on a 27-step horizon for the 3D generalized Hénon map, with $m \in [1, 6]$

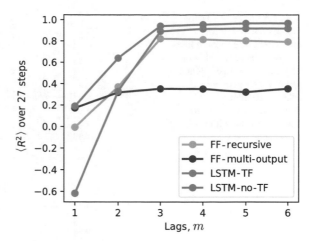

and overestimated ($m > 3$). Comparing Figs. 5.4 and 5.5, we can see that, while the numerical imprecision caused by useless inputs is almost negligible in one step (note again the logarithmic scale in Fig. 5.4), it becomes more significant when considering the whole forecasting horizon (27 steps in Fig. 5.5), also showing considerable differences between predictors. The FF-multi-output is definitely the least robust predictor. The FF-recursive is fairly robust, though it suffers the chaotic nature of the data. LSTM structures are the most robust to the overestimation of the number of necessary lags. Note that the average R^2 score even increases by exploiting the redundant inputs. Without TF, the predictor is also robust to the underestimation of m, confirming that TF should not be used when dealing with forecasting tasks.

5.2   Stochastic Time Series

The results presented in the previous section are relative to noise-free time series, an ideal condition that is never verified when considering practical applications. To assess the sensitivity of the predictors to the observation noise, we superimpose additive white Gaussian noise to the signals obtained by simulating the deterministic systems. We then retrain the four neural predictors on the noisy data. This stochastic environment is thus somehow in the middle between a fully deterministic environment and a real-world case study.

Figure 5.6 reports the results obtained when the white noise standard deviation is equal to the 1% of the noise-free process standard deviation.

As expected, the performances are considerably worse than those obtained in the noise-free case (Fig. 5.1). This is due to the chaotic behavior of the considered systems which exponentially amplify the noise on the initial condition [16]. FF-recursive and LSTM-TF predictors exhibit nearly the same performance in the four systems. Their accuracy is high in the initial part of the forecasting horizon, and degrades to -1 (see

Fig. 5.6 $R^2(\mathbf{y}, \hat{\mathbf{y}}^{(i)})$ score obtained with the four predictors (see colored lines) on the logistic (**a**), Hénon (**b**), 3D generalized Hénon (**c**) and 10D generalized Hénon (**d**) maps. Performances computed on the test dataset with a 1% noise level

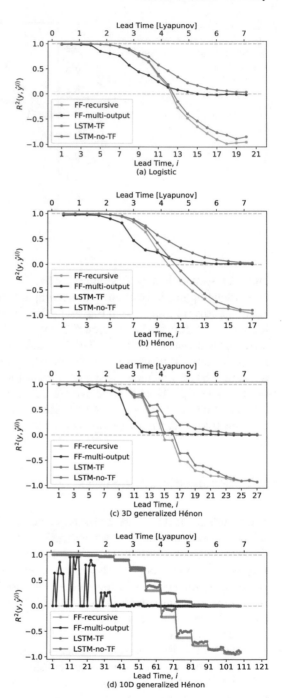

Sect. 4.2). FF-multi-output confirms the peculiar behavior shown in the noise-free case. It ensures a positive (at worst null) R^2 score, providing acceptable performances only when the multi-step forecasting horizon is short. LSTM-no-TF shows the same performance of FF-recursive and LSTM-TF in the first part of the horizon, its R^2 score is always positive as the FF-multi-output (see the last part of the horizon) and it outperforms the three competitors in the central part of the horizon.

An interesting benchmark for the neural architectures is the performance obtained with the real systems generating the data used as predictors (real-system predictor hereafter). In the noise-free case, a real-system predictor always provides a perfect prediction; R^2 score is thus equal to 1 in the whole forecasting horizon. When considering noisy data, we are simulating the same dynamical system from two slightly different initial conditions. The two trajectories diverge due to the systems' chaoticity, as reported in Fig. 5.7. In other words, even if one would be able to identify the real system, it is still not possible to prevent the multi-step error divergence.

We analyze three different levels of noise: its standard deviation is set to 0.5, 1 and 5% of the standard deviation of the noise-free process. As expected, the divergence between real systems and real-system predictors is faster when the noise level is higher. Looking across the different panels, it is clear that the decreasing trends are quite similar for all the chaotic systems considered. This fact is interesting since it proves that adopting the system's Lyapunov time as temporal unit (x axis) and the R^2 score as metric (y axis) is an appropriate way to standardize the dimension of the problem.

The comparison between Fig. 5.6 and the corresponding curve in Fig. 5.7 (the one obtained for 1% noise level) allows us to evaluate the two components of the prediction error separately. The first is caused by the uncertainty in the identification process (identification error). The second is due to the propagation of the observation noise from the input data to the output (observation error). The neural predictors are affected by both sources of error, while the identification error is by definition null when considering the real system used as a predictor. The fact that the FF-recursive and LSTM-TF architectures show a predictive power similar to the real system used as a predictor confirms that these two neural predictors essentially solve a system identification task (they are optimized 1-step-ahead) rather than searching for the best multiple-step forecasting. Conversely, the LSTM-no-TF is specific for the considered horizon and, due to the way it is optimized, it provides better performance than the other competitors. We can conclude that the identification error is almost negligible compared to the observation error in forecasting tasks. This means that the knowledge of the actual system does not help to improve the predictive accuracy when dealing with a chaotic dataset affected by observation noise.

Table 5.3 reports the sensitivity analysis performed by testing three noise levels in terms of number of LTs for which the R^2 score is higher than 0.5.

Table 5.3 confirms that the FF-recursive and LSTM-TF predictors provide similar performances to those obtained by the real system. LSTM-no-TF has a better forecasting accuracy than all the other predictors in all the considered cases.

Fig. 5.7 $R^2(\mathbf{y}, \hat{\mathbf{y}}^{(i)})$ score obtained using the real systems as predictors for different noise levels (see grey scale lines) on the logistic (**a**), Hénon (**b**), 3D generalized Hénon (**c**) and 10D generalized Hénon (**d**) maps

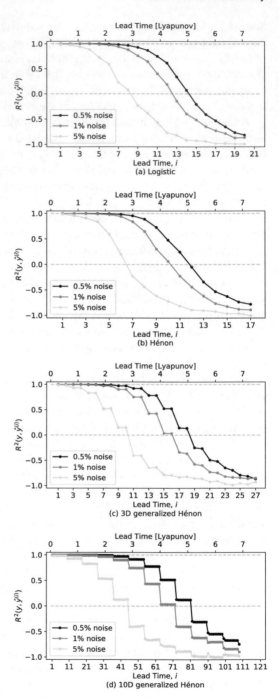

Table 5.3 Performance of the predictors with different levels of noise

System	Predictor	# LT $R^2 > 0.5$			
		Noise free	0.5% noise	1% noise	5% noise
Logistic	FF-recursive	6.37	4.25	3.54	2.12
	FF-multi-output	3.18	2.83	2.83	2.12
	LSTM-TF	> 7.07	4.25	3.54	2.12
	LSTM-no-TF	> 7.07	4.60	3.89	2.12
	Real system	∞	4.24	3.53	2.12
Hénon	FF-recursive	6.43	3.86	3.43	2.14
	FF-multi-output	3.00	3.00	2.57	2.14
	LSTM-TF	> 7.28	3.86	3.43	2.14
	LSTM-no-TF	> 7.28	4.29	3.86	2.57
	Real system	∞	3.86	3.43	2.16
3D generalized Hénon	FF-recursive	6.07	3.86	3.31	2.21
	FF-multi-output	2.76	2.76	2.48	2.21
	LSTM-TF	6.62	4.42	3.31	2.21
	LSTM-no-TF	> 7.45	4.42	3.86	2.21
	Real system	∞	4.42	3.31	2.21
10D generalized Hénon	FF-recursive	6.23	3.97	3.40	2.27
	FF-multi-output	0.00	0.01	0.00	0.01
	LSTM-TF	6.23	3.97	3.40	2.27
	LSTM-no-TF	6.80	4.54	3.97	2.27
	Real system	∞	4.54	3.40	2.27

5.3 Non-Stationary System

After considering the effect of the observation noise, we analyze the results obtained with the slow-fast version of the logistic map to evaluate the effect of the structural noise on the predictors' performance. In this case, the issue is due to the fact that the parabola which defines the mapping between $y(t)$ and $y(t + 1)$ slowly changes with time.

The results obtained for different values of m (i.e., the number of lags fed as an input to the predictor) are reported in Fig. 5.8. As expected, when d is equal to 1, the information is not sufficient to have a mapping between input space and output space. Figure 5.8a shows that none of the predictors can reach an R^2 score close to one even on the one-step-ahead prediction. For m greater than one (Fig. 5.8b–d), the accuracy of all the predictors sensibly increases. The single-step forecasting is

Fig. 5.8 $R^2(\mathbf{y}, \hat{\mathbf{y}}^{(i)})$ score obtained with the four predictors (see colored lines) on the slow-fast logistic map for $m = 1$ (**a**), $m = 2$ (**b**), $m = 3$ (**c**) and $m = 10$ **d**. Performances computed on the test dataset

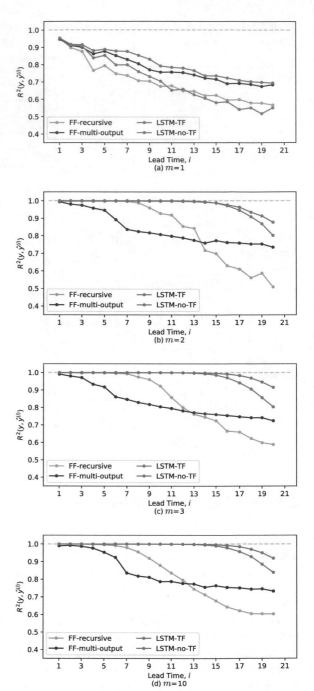

almost perfect meaning that $m = 2$ is an appropriate embedding dimension for this system.

In general, the recurrent structure of the LSTM nets provides better predictive accuracy than a feed-forward one (FF-recursive and FF-multi-output). This is probably due to the fact that the LSTM architectures have an internal memory and their parameters' values change after processing each input. Such networks are thus better at reproducing the slow-varying dynamic of the parameter $r(t)$. Training without teacher forcing further improves the performance, proving that LSTM-no-TF is the more accurate architecture among those considered in this book.

5.4 Real-World Study Cases

In the last section of this chapter, we focus on the results obtained for the two real-world datasets: solar irradiance and ozone concentration. The performance of the predictors is evaluated on a 12-h and a 2-day forecasting horizon, considering in both cases an hourly time step.

Because of the intrinsic noisy nature of real-world time series—where both observation and structural sources of noise typically coexist—before addressing the time series' complexity in terms of its chaoticity, one should question whether a deterministic or stochastic model is more appropriate for describing the series.

This is related to the concept of embedding dimension for the series, introduced in Subsect. 2.4.1. If a suitable embedding dimension is found (e.g., using the false nearest neighbor algorithm), the dynamics in the space of delayed coordinates is uniquely determined (up to the algorithm tolerances), so that a deterministic machine can be used as a model. Conversely, if a significant fraction of false neighbors remains present up to high dimensions, a stochastic model could be more appropriate. Indeed, in such cases, the correlation dimension (Subsect. 2.3.2) of the reconstructed attractor keeps increasing while enlarging the dimension of the embedding space, a sign that the stochastic nature of the data fills the available space.

Once a suitable embedding is found, the time series complexity can be measured by computing the largest Lyapunov exponent (LLE) of the reconstructed attractor (Subsect. 2.4.2). A positive exponent certifies the time series as chaotic, i.e., exponentially sensitive to small perturbations.

5.4.1 Solar Irradiance

The procedure described above first requires the evaluation of a suitable embedding dimension for the dataset, using the false nearest neighbor algorithm. The fraction of false neighbors is almost null for $m = 48$ (see Fig. 5.9a), meaning that such a value of the delay parameter is appropriate for reconstructing the system dynamic.

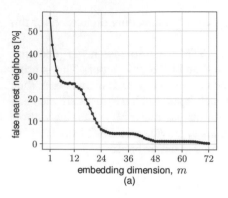
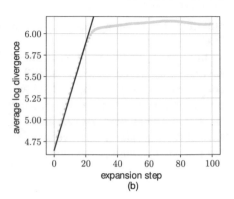

Fig. 5.9 Output of the false nearest neighbor algorithm (**a**) and estimation of the LLE (**b**) for the solar radiation time series (Como, 2019)

After having set the value of m, we compute the average logarithmic divergence for an increasing expansion step. From the resulting curve, reported in Fig. 5.9b, it is possible to compute the LLE, i.e., the slope of the initial (linear) part of the curve. In this case, it is equal to 0.062.

Since the LLE associated to this time series is positive, we can confirm that the solar irradiance dynamics is chaotic. The same conclusion has been reached in a study by Fortuna et al. [4] in which the authors analyzed the solar irradiance recorded in the same station considered in this book.

The comparison of the multi-step prediction of solar irradiance with LSTM and FF networks is performed using data from 2014 to 2017 for networks' parameters training, 2018 for the validation process and 2019 to test the predictors fairly.

Before presenting the results, it is worth noting that the classical indicators may overestimate the actual performances of models when applied to the complete time series. When dealing with solar irradiance, there is always a strong bias due to the presence of many null values. In the case at hand, they are about 57% of the sample due to different factors: the rotation and revolution motion of the Earth, the additional shadowing of the nearby mountains and the sensitivity of the sensors. When the recorded value is zero, the forecast is also zero (or very close) and all these small errors substantially reduce the average error and increase the R^2 score. Additionally, forecasting the solar irradiance during the night is useless, and the power network manager that uses the forecast to balance electricity production may well turn the forecasting model off. In order to overcome this deficiency, which unfortunately is present in many works in the current literature, and allow the models' performances to be compared when they may indeed be useful, we have also computed the performance indicators considering only values above 25 Wm^{-2} (daytime in what follows), a small value normally reached before dawn and after sunset. These are the conditions when an accurate solar energy forecast may turn out to be useful.

First, we consider the performance obtained by two standard baseline models: the clear sky index and a classical persistent model. The clear sky model (see Fig. 3.6b) inherently takes into account the presence of the daily pseudo-periodic component, which affects hourly global solar radiation, and considers average clear sky conditions. The clear sky predictor preserves the same performances for each step ahead since they are independent of the considered lead time: the R^2 score is equal to 0.58 during the whole day and 0.22 when focusing on daytime samples only. The persistent predictor is defined as $\hat{y}(t + i) = y(t), i = 1, 2, \ldots, h$. It thus takes into account the very short-term memory of the process. Its performance rapidly deteriorate when increasing the horizon. It is acceptable 1 hours ahead; the R^2 score is equal to 0.79 (considering whole day samples) and 0.59 (daytime samples only). After two hours, it decreases to 0.54 (whole day) and 0.14 (daytime only). Further increasing the lead time, the performance dramatically drops; for instance, the R^2 score becomes -0.93 (whole day) and -1.64 (daytime) after six hours.

Figure 5.10 reports the results obtained with the four neural predictors in terms of the R^2 score both during the whole day and when focusing on daytime samples only.

As expected, the neural predictors outperform the two benchmarks. They provide better performances than the persistent model in the short term and better than the clear sky index on the whole horizon (both considering $h = 12$ and $h = 48$). The only exception is represented by the LSTM-TF, whose performance over the whole day is worse than the one provided by the clear sky model after 24 steps ahead (see, row 3, column 2, in Fig. 5.10).

Figure 5.10 shows that the LSTM-no-TF predictor provides the best performance. Comparing the latter with the FF-multi-output predictor, it emerges that the two have a similar accuracy on the 12-step horizon, while FF-multi-output predictor performs slightly worse after 24 steps ahead (when considering the 48-step horizon). Adopting the FF-recursive approach, the R^2 score decreases, specifically after 5 hours ahead. Finally, considering the LSTM-TF predictor, the performance drops dramatically. This is probably due to the highly periodic trend characterizing the solar irradiance, which requires the proper propagation of the information through the internal state of the recurrent cells. The LSTM architecture fails in such a task if it is trained with the traditional teacher forcing procedure. For all the considered neural predictors, the difference between the whole time series (average value 140.37 Wm^{-2}) and the case with a threshold (daytime only), which excludes nighttime values (average 328.62 Wm^{-2}), emerges clearly, given that during all nights the values are zero or close to it and can be easily predicted, and thus the corresponding errors are also low. Therefore, removing the nighttime values from the test set of the time series is crucial for a realistic assessment of a solar radiation forecasting model, which would otherwise be strongly biased.

An additional way to examine the predictors' performances is presented in Table 5.4. We report the R^2 score of the predictions of the LSTM network on three horizons, namely 1, 3 and 6 hours ahead. The sunlight (i.e., above 25 W/m^2) test series is partitioned into three classes: cloudy, partly cloudy and sunny days, which constitute about 30, 30 and 40% of the sample, respectively. More precisely, cloudy days are defined as those when the daily average irradiance is below 60% of the clear

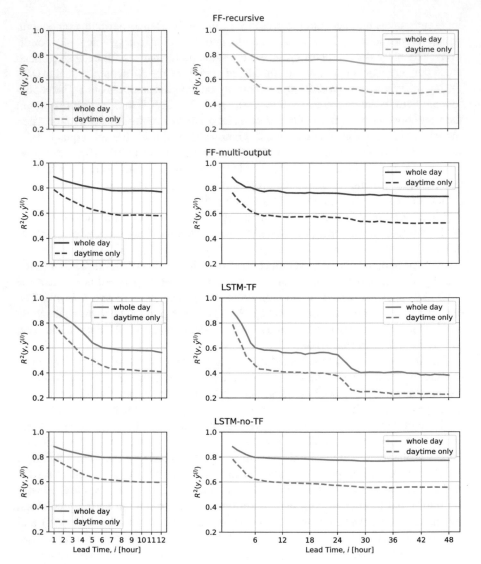

Fig. 5.10 $R^2\left(\mathbf{y}, \hat{\mathbf{y}}^{(i)}\right)$ score for the four predictors (see colored lines) on the solar irradiance time series. Solid lines represent the performance during the whole day, dashed lines during daytime only. Performances computed on the test dataset (Como, 2019) for $h = 12$ (first column), and $h = 48$ (second column)

Table 5.4   LSTM-no-TF performance in terms of R^2 score for cloudy, partly cloudy and sunny days (daytime samples only)

Weather condition	1 hours ahead	3 hours ahead	6 hours ahead
Cloudy	0.44	0.06	−0.45
Partly cloudy	0.65	0.59	0.59
Sunny	0.89	0.83	0.73

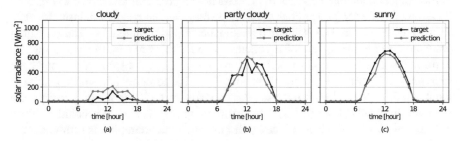

Fig. 5.11   Three-hour-ahead LSTM-no-TF predictions versus observations in three individual days with different cloudiness: a cloudy day (**a**), a partly cloudy day (**b**) and a sunny day (**c**). Values refer to the test year (2019)

sky index and sunny days are those that are above 90% (remember that the clear sky index already accounts for the average sky cloudiness).

It is quite apparent that the performance of the model decreases consistently from sunny, to partly cloudy, to cloudy days. This result is better illustrated in Fig. 5.11 where the 3-hours-ahead predictions are shown for three individual days. In the sunny day, on the right, the process is almost deterministic (governed mainly by astronomical conditions), while the situation is completely different on a cloudy day. In the last case, the forecasting error is of the same order of the process (R^2 score close to zero) and it can be even larger at 6 hours ahead. This determines the negative R^2 score value shown in Table 5.4.

Besides the accuracy of the forecasted values, another important characteristic of the predictive models is their generalization capability, often mentioned as domain adaptation in the neural networks literature [5]. This means the possibility of storing knowledge gained while solving one problem and applying it to different, though similar, datasets [19]. To test this feature, the FF and LSTM networks developed for the Como station (source domain) have been used, without retraining, on the sites represented in Fig. 3.7 (target domains) which have substantially different geographical conditions.

In addition, the test year has been changed because solar radiation is far from being truly periodical and some years (e.g., 2017) show significantly higher values than others (e.g., 2011): this means quite different solar radiation encompassing a difference of about 25% between yearly average values. Figure 5.12 (first three rows) shows the R^2 score for the LSTM-no-TF networks for three additional stations: Casatenovo (test year 2011), Bigarello (test year 2016) and Bema (test year 2017).

All the graphs reach a plateau after 6-8 steps ahead and the differences between FF and LSTM networks appear very small or even negligible in almost all the other stations. Six hours ahead, the difference in R^2 score between Como, for which the networks have been trained, in the test year (2019) and Bema in 2017, which appears to be the most different dataset, is only about 3% for both FF models and LSTM.

As a further trial, both FF models and LSTM have been tested on a slightly different process, i.e., the hourly average solar radiation recorded at the Como station. Forecasting results obtained with the LSTM-no-TF predictor are shown in Fig. 5.12 (last row). While the process has the same average as the original dataset, its variability is different since its standard deviation decreased by about 5%. The averaging process indeed filters the high frequencies. Also for this process, the neural models perform more or less the same as they do for the hourly values for which they have been trained. For a correct comparison with Fig. 5.10, however, it is worth bearing in mind that the 1-hour ahead prediction corresponds to $(t + 2)$ in the graph, since the average computed at hour $(t + 1)$ is that from (t) to $(t + 1)$ and thus includes values that are only 5 minutes ahead of the instant at which the prediction is formulated.

An ad-hoc training on each sequence would undoubtedly improve the performance, but the exact purpose of this analysis is to show the potential of networks calibrated on different stations, to evaluate the possibility of adopting a predictor developed elsewhere when a sufficiently long series of values is missing. The forecasting models we developed for a specific site could be used with acceptable accuracy for sites where recording stations are not available or existing time series are not long enough. This suggests the possibility of developing a unique forecasting model for the entire region wherever the solar radiation process is not influenced by other external factors.

The results obtained on the solar irradiance dataset (both source and target domains) are summarized in Table 5.5.

5.4.2 Ozone Concentration

To check whether the ozone concentration time series can be considered as chaotic, we repeat the same analysis previously performed for the solar irradiance.

As shown in Fig. 5.13a, the fraction of false neighbors is almost zero for an embedding dimension equal to 24 steps ($m = 24$). This is, therefore, an appropriate number of autoregressive terms to be taken into account to reconstruct the ozone dynamic. In the delayed coordinate space, we analyze the average logarithmic divergence (see Fig. 5.13b), from which we compute the slope of the linear part of the curve. The resulting LLE is equal to 0.057.

Other authors performed similar analyses on ozone time series in different locations: Cincinnati, Ohio [2], Arosa, Switzerland [1], and Berlin, Germany [7]. Their results are in accordance with ours and indicate that the ozone dynamics can be characterized as a chaotic system.

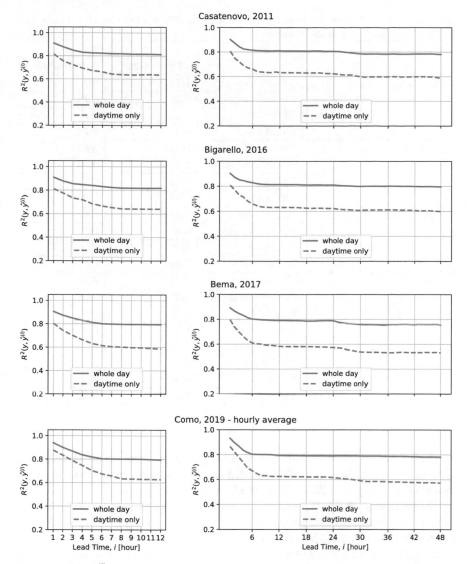

Fig. 5.12 $R^2(\mathbf{y}, \hat{\mathbf{y}}^{(i)})$ score obtained with the LSTM-no-TF predictor on the solar irradiance time series. Domain adaptation tested on different sites in the Lombardy region for $h = 12$ (first column), and $h = 48$ (second column). Solid lines represent the performance measured during the whole day, dashed lines during daytime only

Table 5.5 Overall performance of the predictors over the different solar irradiance time series considered. R^2 scores computed in whole day

Dataset	Predictor	$\langle R^2 \rangle$	
		12 steps ahead	48 steps ahead
Como, 2019	FF-recursive	0.79	0.75
	FF-multi-output	0.80	0.76
	LSTM-TF	0.66	0.51
	LSTM-no-TF	0.81	0.78
Casatenovo, 2011	FF-recursive	0.83	0.80
	FF-multi-output	0.83	0.80
	LSTM-TF	0.68	0.53
	LSTM-no-TF	0.83	0.80
Bigarello, 2016	FF-recursive	0.84	0.81
	FF-multi-output	0.84	0.81
	LSTM-TF	0.70	0.54
	LSTM-no-TF	0.84	0.81
Bema, 2017	FF-recursive	0.81	0.76
	FF-multi-output	0.82	0.78
	LSTM-TF	0.68	0.54
	LSTM-no-TF	0.82	0.78
Como, 2019 hourly average	FF-recursive	0.83	0.78
	FF-multi-output	0.83	0.78
	LSTM-TF	0.64	0.52
	LSTM-no-TF	0.83	0.80

The training of the four multi-step predictors is performed on the dataset from 2008 to 2013. The data from 2014 to 2015 are used for the validation process and the data from the period 2016–17 are used for the testing phase.

The persistent model [6, 11, 14], $\hat{y}(t + i) = y(t)$, $i = 1, 2, \ldots, h$, used as a benchmark, provides quite a good performance in the very short term but its accuracy rapidly deteriorates as the lead time increases. After 1 hour, the R^2 score is equal to 0.90, after two hours it decreases to 0.78, after three hours to 0.63. Further increasing the lead time, the predictive accuracy drops to low values; for instance, the R^2 score is 0.26 after six hours.

The results for $h = 12$ and $h = 48$ are presented in the first and second column of Fig. 5.14. Considering the 12-hour horizon, the four predictors provide similar performances. A relevant gap emerges in the second half of the forecasting horizon between the LSTM-no-TF and the FF-recursive predictors: the 12-step ahead R^2 score is 0.60 for the former and 0.49 for the latter. FF-multi-output and LSTM-TF ensure intermediate performances.

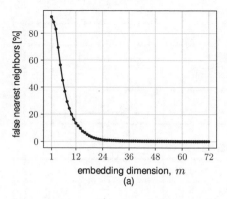
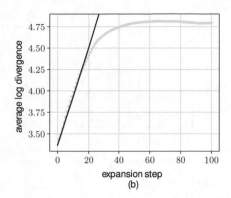

(a)   (b)

Fig. 5.13 Output of the false nearest neighbor algorithm (**a**) and estimation of the LLE (**b**) for the ozone concentration time series (Chiavenna, 2016-17)

Table 5.6 Overall performance of the predictors computed on the ozone concentration time series

Dataset	Predictor	$\langle R^2 \rangle$	
		12 steps ahead	48 steps ahead
Chiavenna, 2016–2017	FF-recursive	0.66	0.39
	FF-multi-output	0.70	0.49
	LSTM-TF	0.70	0.45
	LSTM-no-TF	0.73	0.53

By increasing the length of the forecasting horizon to 48 steps (second column of Fig. 5.14), we spot the same ranking: LSTM-no-TF ensures the best performance, followed by FF-multi-output, LSTM-TF and FF-recursive. In particular, it emerges that for a long forecasting horizon, it is important to optimize the predictor on the whole horizon (LSTM-no-TF and FF-multi-output) rather than on the single step (FF-recursive and LSTM-TF). In the first case, the R^2 score tends to 0.4, while in the second it tends to 0.2. The difference is thus remarkable, even if in all cases the R^2 scores are under 0.6, which is traditionally considered as the lower threshold below which a model has an insufficient precision to be of use in practical applications.

From the qualitative point of view, all the predictors have a common trend. The performance is almost flat between 18 and 24 hours ahead and then decreases again, reaching another plateau at the end of the horizon. This behavior could be due to the fact that the ozone concentration dynamic on a given day is somewhat decoupled from the one of the following day due to the absence of the ultraviolet radiation coming from the sun during nighttime.

The overall performances obtained for the ozone concentration dataset on the 12-step ahead and on the 48-step ahead horizon are reported in Table 5.6.

Figure 5.15 shows the trajectories predicted by the four neural nets in a specific case (the target is the same in the whole panel). As expected, all the predictors have

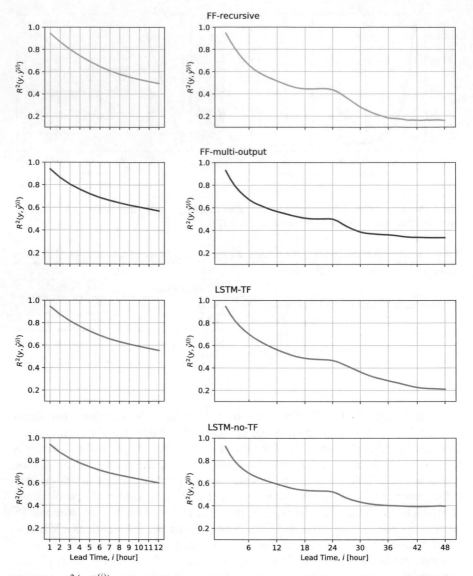

Fig. 5.14 $R^2(\mathbf{y}, \hat{\mathbf{y}}^{(i)})$ score for the four predictors on the ozone concentration time series. Performances computed on the test dataset (Chiavenna, 2016–2017) for $h = 12$ (first column) and $h = 48$ (second column)

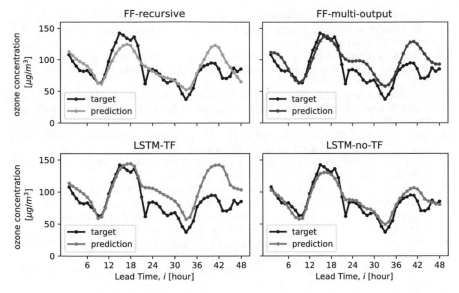

Fig. 5.15 Target sequence and corresponding predictions for a specific 2-day horizon ($h = 48$) from the ozone concentration test set

a better performance in the first part of the horizon and struggle with the forecasting of the final part. The comparison reported in Fig. 5.15 also shows that the prediction sequences are smoother than the observed data. This is proof that the predictor is not modelling the noise that affects this real-world dataset.

References

1. Chattopadhyay, G., & Chattopadhyay, S. (2008). A probe into the chaotic nature of total ozone time series by correlation dimension method. *Soft computing, 12.10*, 1007–1012.
2. Chen, J. L., Islam, S., & Biswas, P. (1998). Nonlinear dynamics of hourly ozone concentrations: Nonparametric short term prediction. *Atmospheric environment, 32.11*, pp. 1839–1848.
3. Dercole, F., Sangiorgio, M., & Schmirander, Y. (2020). An empirical assessment of the universality of ANNs to predict oscillatory time series. *IFAC-PapersOnLine, 53.2*, 1255–1260.
4. Fortuna, L., Nunnari, G., & Nunnari, S. (2016). *Nonlinear modeling of solar radiation and wind speed time series*. Springer.
5. Glorot, X., Bordes, A., & Bengio, Y. (2011). Domain adaptation for large-scale sentiment classification: A deep learning approach. In *Proceedings of the 28 th International Conference on Machine Learning*, Bellevue, WA, USA.
6. Guariso, G., Nunnari, G., Sangiorgio, M. (2020). Multi-Step Solar irradiance forecasting and domain adaptation of deep neural networks. *Energies, 13.15*, 3987.
7. Haase, P., Schlink, U., & Richter, M. (2002). Non-parametric short-term prediction of ozone concentration in Berlin: Preconditions and justification. In *Air Pollution Modelling and Simulation* (pp. 527–536). Springer.
8. Kennel, M. B., Brown, R., & Abarbanel, H. D. (1992). Determining embedding dimension for phase-space reconstruction using a geometrical construction. *Physical Review, A, 45.6*, 3403.

9. Manabe, Y., & Chakraborty, B. (2007). A novel approach for estimation of optimal embedding parameters of nonlinear time series by structural learning of neural network. *Neurocomputing, 70.7-9*, 1360–1371.
10. Maus, A., & Sprott, J. C. (2011). Neural network method for determining embedding dimension of a time series. *Communications in nonlinear science and numerical simulation*, 16.8, 3294–3302.
11. Meyer, P. G., Kantz, H., & Zhou, Y. (2021). Characterizing variability and predictability for air pollutants with stochastic models. *Chaos: An Interdisciplinary Journal of Nonlinear Science, 31.3*, 033148.
12. Ouala, S., et al. (2020). Learning latent dynamics for partially observed chaotic systems. *Chaos: An Interdisciplinary Journal of Nonlinear Science, 30.10*, 103121.
13. R. J. Povinelli, et al., Statistical models of reconstructed phase spaces for signal classification. *IEEE Transactions on Signal Processing, 54.6*, 2178–2186.
14. Sangiorgio, M. (2021). Deep learning in multi-step forecasting of chaotic dynamics. Ph.D. thesis. Department of Electronics, Information and Bioengineering, Politecnico di Milano.
15. Sangiorgio, M., & Dercole, F. (2020). Robustness of LSTM neural networks for multi-step forecasting of chaotic time series. *Chaos, Solitons & Fractals, 139*, 110045.
16. Sangiorgio, M., Dercole, F., & Guariso, G. (2021). Sensitivity of chaotic dynamics prediction to observation noise. *IFAC-PapersOnLine*, 54.17, 129–134.
17. Takens, F. (1981). Detecting strange attractors in turbulence. In *Dynamical systems and turbulence, Warwick 1980*, pp. 366–381. Springer.
18. Yijie, W., Min, H. (2007). Prediction of multivariate chaotic time series based on optimized phase space reconstruction. In *Proceedings of the Chinese Control Conference* (pp. 169–173).
19. Yosinski, J., et al. (2014). How transferable are features in deep neural networks? In *Proceedings of the 28th Conference on Neural Information Processing Systems*, 27, 3320–3328.

Chapter 6
Neural Predictors' Sensitivity and Robustness

Abstract The results of the application of deep neural predictors depend on a multitude of factors which compose the experimental settings. We report all the specific information to ensure the reproducibility of a wide number of numerical experiments. A sensitivity analysis on some critical aspects is provided in order to prove the robustness of our setting. Considering the long-term behavior of the predictors, those trained for the one-step forecasting are able to reproduce the statistical properties of the attractor, i.e., the so-called attractor's climate, whereas the multi-step ones are unsuitable for replicating these statistical properties but provide an accurate forecasting up to several Lyapunov times. Lastly, we provide some remarks on the training procedure of the different predictors and introduce some advanced neural architectures to give an overview of possible advantages/disadvantages with respect to those implemented in this study.

6.1 Simplicity and Robustness of the Experimental Setting

An exhaustive analysis of all the aspects of the problem (i.e., the class of chaotic systems, the number of samples and the features of the neural architectures) would require an unmanageable computational effort. We thus fix these elements in order to avoid choices which could potentially affect the results.

First, we avoid continuous-time systems because numerical integration unavoidably introduces errors, especially in chaotic systems, so that the neural network ends up learning the discrete-time representation of the system defined by the integrator. Besides numerical integration, which can be accurately performed, the choice of the sampling time strongly affects the results. A short sampling time makes the single-step predictor almost equal to the identity map. This dramatically impairs the performance of FF-recursive and LSTM-TF predictors. Moreover, many steps are required for each Lyapunov time, preventing FF-multi-output predictors from reaching acceptable performances (an analogous situation is represented in Fig. 5.1d). On the other hand, a long sampling time makes one-step mapping too complex (see Fig. 3.2). In addition, we use systems that can be explicitly formulated as a

Table 6.1 Number of parameters (weights and biases) of the considered neural predictors. The values are expressed as function of m (number of lags), h (number of leads) and n_n (number of neurons)

Layer	FF-recursive	FF-multi-output	LSTM
Hidden #1	$n_n \cdot (m + 1)$	$n_n \cdot (m + 1)$	$4 \cdot n_n \cdot (n_n + 3)$
Hidden #2	$n_n \cdot (n_n + 1)$	$n_n \cdot (n_n + 1)$	$8 \cdot n_n \cdot (n_n + 1)$
Hidden #3	$n_n \cdot (n_n + 1)$	$n_n \cdot (n_n + 1)$	$n_n \cdot (n_n + 1)$
Output	$n_n + 1$	$h \cdot (n_n + 1)$	$n_n + 1$

multi-step recurrence on a single output variable $y(t)$. For these systems, the dimension of the recurrence gives the number of inputs m (lags) that makes the forecasting problem feasible (the embedding dimension of the dataset), hence avoiding the problems related to its numerical determination.

Second, we fix the number of hidden layers (3), each with a fixed number of neurons ($n_n = 10$). Note that for FF architectures, the number of the network's parameters still depends on the number of inputs (the m lags of the system to be learned) and outputs (the predictive horizon length h), which are system-specific. Table 6.1 reports the number of parameters for each neural architecture. The number of parameters of the FF-recursive and the FF-multi-output are the same for the hidden layers. The only difference between them is in the output layer. The FF-recursive predictor computes only one step, while the FF-multi-output predictor returns the whole h-step output sequence. LSTM-TF and LSTM-no-TF are reported together (LSTM column) since they have exactly the same architecture (the two predictors differ for the training procedure).

The number of weights and biases of LSTM nets displayed in Table 6.1 needs to be discussed briefly. First, it does not depend on m and h because of the parameters shared across different time steps. Second, the default PyTorch implementation of the recurrent neurons has two biases for each of the four nonlinearities represented in the bottom part of the LSTM cell in Fig. 4.3. The two biases are simply summed up, increasing the number of parameters without changing the degrees of freedom. Considering, for instance, the first hidden layer, it has $4n(n + 3)$ parameters even if, in fact, the same degrees of freedom can be obtained with $4n(n + 2)$ by removing the redundant biases. We performed some further experiments that show how this issue does not affect the results.

An alternative to use a constant number of hidden layers could be to fix the number of network parameters. In this case, the same number of parameters can be obtained for many different combinations (# of neurons per layer, # of hidden layers) and thus an arbitrary choice is still necessary. After considering several combinations, we selected the one which ensures a reasonable compromise between performance and training convergence.

In LSTM predictors, an additional arbitrary choice about the number of LSTM and FF layers is necessary. The results presented in the previous chapter are obtained with two LSTM layers followed by an additional fully-connected hidden layer. Alternatively, one can decide to use a single LSTM layer with two FF hidden layers on top of it. The comparison between the two configurations, trained with (Fig. 6.1a) and without (Fig. 6.1b) TF, shows that they have almost equivalent performances. A single LSTM layer is thus sufficient to capture the time-varying dynamics of the data and to keep an internal state of the processed sequence.

We also maintain a constant number of samples (50,000) for the artificial systems. As we have underlined while presenting our results, we do it intentionally, in connection with the choice of the chaotic data generators, to challenge our predictors on tasks with different complexity. On the other hand, scaling the datasets' and network's dimensions with the complexity of the task (in our case a good proxy would be the fractal dimension of the chaotic attractor) would be useful to compare the performance of each predictor across different systems [6].

To further support the robustness of our main result—performance ranking among the four predictors on each of the considered forecasting tasks—we confirm that we always observed the ranking of Fig. 5.1 in all unreported tests that we have run before deciding to fix the experimental setup as described here.

6.2 Predictors' Long-Term Behavior

The main focus of this book is on the multi-step forecasting: the predictors have been evaluated on the output sequence $[y(t + 1), y(t + 2), \ldots, y(t + h)]$, specifically considering a fixed-term horizon. A related but different task is the system identification, which consists of searching for the best copy of the system. Following this terminology, we can say that FF-multi-output and LSTM-no-TF are trained to be optimal multi-step predictors. Conversely, FF-recursive and LSTM-TF are optimized to be the best copies of the system: they try to reproduce the relationship between the inputs $[y(t), y(t - 1), \ldots, y(t - m + 1)]$ and the one-step value $y(t + 1)$.

In this section, we analyze the predictors' long-term behavior, i.e., their ability to reproduce the "climate" of the chaotic attractor [2, 22]. Note that the FF-multi-output predictor has been excluded from the long-term analysis because it cannot be used as a dynamical system to generate a sequence of output of arbitrary length due to its static nature.

Figure 6.2 provides exactly the same information of Fig. 5.1, but for a longer forecasting horizon. The latter focuses on the specific horizon for which the predictors are trained (see Table 5.1), while in the former, the number of steps ahead is increased to show the regime behavior of the predictors.

Figure 6.2 confirms that FF-recursive and LSTM-TF reproduce the chaotic attractor of the original system well, since their R^2 score goes to -1 (see Sect. 4.2). This is not the case with LSTM-no-TF, whose behavior beyond the forecasting horizon is not guaranteed [25].

An alternative perspective to look at the same issue is the analysis of the LLEs of the predictors, reported in Table 6.2.

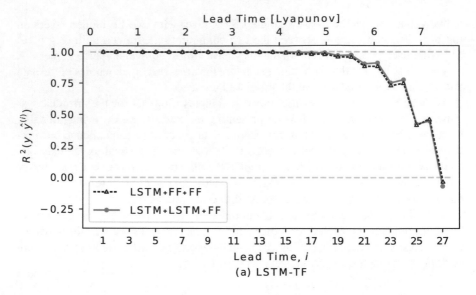

(a) LSTM-TF

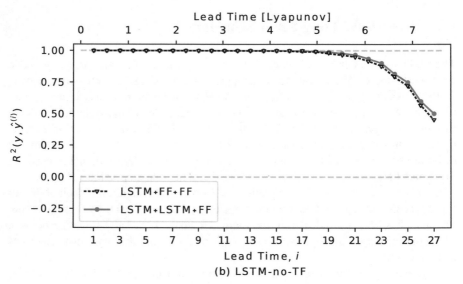

(b) LSTM-no-TF

Fig. 6.1 $R^2(\mathbf{y}, \hat{\mathbf{y}}^{(i)})$ score comparing different LSTM architectures (two LSTM layers with a FF layer on top, and a single LSTM layer with two FF layers on top) trained with (**a**) and without (**b**) TF. Performances computed on the 3D generalized Hénon map with the noise-free test dataset

Fig. 6.2 Long-term $R^2\left(\mathbf{y}, \hat{\mathbf{y}}^{(i)}\right)$ score of FF-recursive, LSTM-TF, and LSTM-no-TF (see colored lines) on the logistic (**a**), Hénon (**b**), 3D generalized Hénon (**c**) and 10D generalized Hénon (**d**) maps. Performances computed on the noise-free test dataset

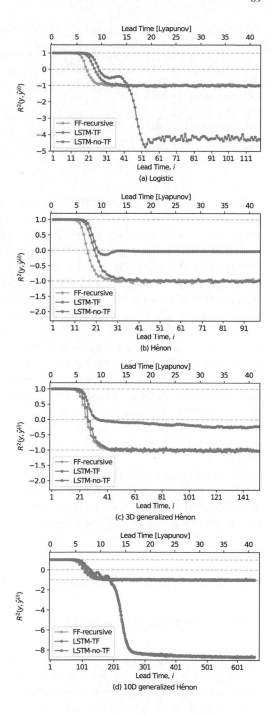

Table 6.2 LLEs of the real system and of the neural predictors

System	Real system	FF-recursive	LSTM-TF	LSTM-no-TF
Logistic	0.354	0.354	0.354	0.354
Hénon	0.432	0.416	0.406	0.126
3D generalized Hénon	0.276	0.280	0.281	0.149
10D generalized Hénon	0.063	0.065	0.065	0.126

The values of the LLEs confirm the findings which emerged from Fig. 6.2. FF-recurrent and LSTM-TF predictors have almost the same LLE of the real system: the percentual error is usually low (it is greater than 5% only in one case).

LSTM-no-TF exhibits different behaviors in the four chaotic systems. It replicates the logistic LLE well (0.354, equal to the actual value), but the long-term R^2 score tends to a value between -4 and -5 (Fig. 6.2a). This means that the predictor's attractor is similar to the original one (same LLE), but it covers a different part of the phase space. The situation is quite different with the Hénon map. The R^2 score approaching 0 (Fig. 6.2b) suggests that the predictor forecasts a value equal to the average of the dataset in the long term. However, the value does not always assume the exact average value as it would seem when looking at the figure. There is a chaotic dynamic (LLE equal to 0.126) covering a small portion of the phase space, which cannot be spotted in the figure scale. In the 3D generalized Hénon map, the situation is quite similar to the one of the Hénon map, with the only difference that long-term dynamics occur (Fig. 6.2c). The 10D generalized Hénon predictor's chaotic attractor has a different LLE from the original one and covers another part of the phase space (Fig. 6.2d).

The results presented in this section confirm that, in general, the best predictor and the best copy of the system do not necessarily coincide. Reproducing the properties of the original chaotic attractor (such as the LEs) is something related to system identification rather than multi-step forecasting.

6.3 Remarks on the Training Procedure

6.3.1 Backpropagation and Backpropagation Through Time

The development of many successful deep learning models in various applications has been made possible by three joint factors. First, the availability of big data, which are necessary to identify complex architectures characterized by thousands of parameters. Second, the intuition of making use of fast parallel processing units (GPUs) able to deal with the high computational effort required. Third, the availability

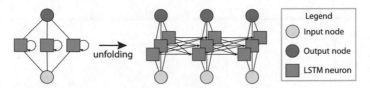

Fig. 6.3 Unfolding of the recurrent LSTM architecture for 3 time steps

of efficient gradient-based methods to train these kinds of neural networks. The latter are the well-known backpropagation (BP) techniques, which allow efficient computing of the gradients of the loss function with respect to each model weight and bias. To apply the backpropagation of the gradient, it is necessary to have a feed-forward architecture (i.e., without self-loops). In this case, the optimization is extremely efficient since the process can be entirely parallelized, exploiting the GPU. When the neural architecture presents some loops, as in recurrent cells, the BP technique has to be slightly modified in order to fit the new situation. This can be done by unfolding the neural networks through time to remove self-loops (Fig. 6.3).

This extension of the BP technique is known as backpropagation through time (BPTT) in the machine learning literature [28]. The issue with BPTT is that the unfolding process should in principle last for an infinite number of steps, making the technique useless for practical purposes. For this reason, it is necessary to limit the number of unfolding steps, considering only those time steps that contain useful information for the prediction (in this case, we say that the BPTT is truncated). As it is easy to understand, BPTT is not as efficient as the traditional BP, because it is not possible to fully parallelize the computation. As a consequence, we are not able to fully exploit the GPU's computing power, resulting in a slower training. The presence of recurrent units also produces more complex optimization problems, with a significant number of local optima [5].

Figure 6.4 shows the substantial differences in the evolution of the training process of the four approaches.

As usual, the mean of the quadratic errors of the FF network smoothly decreases toward a minimum while this function shows sudden jumps followed by several epochs of stationarity in the case of LSTM. The training algorithm of LSTM can avoid being trapped in local minima for too many epochs, but on the other hand, these local minima are much more frequent.

Going into more details, we can draw some interesting remarks on how the training procedure evolves in the four different approaches. The identification of the FF-recursive predictor requires a one-step loss ($\mathrm{MSE}(\mathbf{y}, \hat{\mathbf{y}}^{(1)})$) to be considered. As a consequence, the MSE assumes a low value at the beginning of the training procedure and its trend suggests that the related optimization problem is fairly simple. In the three other cases, the loss function is computed on a h-step horizon: $\frac{1}{h}\sum_{i=1}^{h}\mathrm{MSE}(\mathbf{y}, \hat{\mathbf{y}}^{(i)})$. The LSTM-TF predictor loss starts from a low value due to the teacher forcing procedure (see Fig. 4.4a) but highlights the presence of some local minima which are associated with the BPTT. FF-multi-output and LSTM-no-

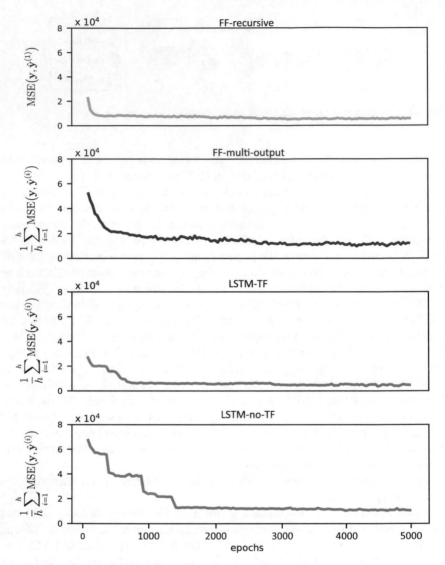

Fig. 6.4 Example of the evolution of the loss function over 5000 training epochs. The four neural predictors are trained on the solar irradiance dataset ($m = 48$, $h = 12$)

TF predictors are actually optimized on the multi-step horizon, as suggested by the high values assumed by the loss function in the first training epochs. Again, the FF-multi-output (whose gradients are computed with the BP) exhibits a smoother trend than that of LSTM-no-TF which presents many local optima.

Another factor that could potentially affect the complexity of the training process is the number of weights and biases to be optimized. The LSTM predictors usually have a higher number of parameters than the FF competitors due to the presence of the connections between internal states which characterize the recurrent architectures. Because of this, the optimization process of LSTM nets is probably more complex (and its convergence slower) than that of FF architectures.

In the case of solar irradiance ($m = 48$, $h = 12$), the FF-recursive architecture has 721 parameters, the FF-multi-output 842, and the two LSTM networks 1521. The number of parameters of the FF predictors is much lower in the other datasets tested (see Table 6.1).

6.3.2 Training with and Without Teacher Forcing

We usually link the concept of training with or without teacher forcing to the specific case of recurrent neural networks. However, the same topic has been widely discussed, under different names, also with reference to FF architectures and, more generally, to empirical identification of dynamical systems and time series prediction [3, 4].

In the case of FF neural networks, the topic has been investigated since the early nineties by referring to series-parallel (analogue to TF) and parallel (no-TF) training modes [12, 13]. In contrast to what happens in NLP, where TF is considered as the standard approach, both TF and no-TF are widely adopted in system identification and time series prediction, with results that are application-specific [17, 24]. The real advantage of TF in FF nets is that, as we have explained in the description of the FF-recursive predictor, it breaks the output-to-input feedback during training and hence transforms a dynamic task into a static one [26]. For this reason, training with TF is amenable to parallelization and thus more efficient under the computational perspective [24]. Conversely, in recurrent nets there are two feedbacks: one from output at the time t, to input at time $t + 1$ as in FF nets, and another in the hidden states. The first can be removed by adopting TF, while the flow of information through the hidden states cannot be teacher forced. As a consequence, even by adopting TF we cannot process each time step as an independent sample as it happens with FF architectures. The two approaches thus require a similar computational effort when training recurrent nets.

Similar results appeared in the field of empirical modelling with polynomial functions. In this context, training with TF shares the main concept with the minimization of the one-step prediction error, while no-TF looks at the performance over a multi-step horizon (simulation error) and is argued as a promising technique for the identification of polynomial models [8–11, 23]. The computational effort required

by the two approaches is, however, significantly different, because the minimization
of the prediction error is simply obtained with the least squares formula.

The use of TF has been widely discussed, also in the context of reservoir computing. In fact, reservoir computers are RNNs trained in a specific way. Instead of
adopting the usual backpropagation almost always used to train neural nets (specifically the backpropagation through time), the recurrent weights are randomly set and
only the parameters of the hidden-to-output linear layer are identified. Under TF, their
optimal value is computed with the least squares formula. Without TF, the peculiar
trick (and advantage) of reservoir computing gets lost, as feeding the prediction at
time $t + k$ as the input for the prediction at time $t + k + 1$ makes the loss a nonlinear
function of the output weights.

6.4   Advanced Feed-Forward Architectures

In recent years, recurrent networks have been losing ground to high-performing
FF architectures (abandoning the traditional fully-connected structure) in several
machine learning tasks.

Special types of FF models that have been successfully adapted to sequence tasks
are convolutional neural networks [1, 19]. 1D convolution through time has the main
role of filtering noise and/or of extracting complex features from the time series (see,
for instance, the seminal paper by LeCun [15], and other recent practical applications [20, 21]). Convolutional layers have also been used in hybrid architectures to
preprocess the signal before feeding it into recurrent layers or, alternatively, on top of
them. Considering the forecasting of deterministic chaotic dynamics, we know that
there is a mapping from input to output and the inputs are noise-free. Convolution
is, therefore, not only unnecessary, it could also have negative effects, distorting the
input and making the reproduction of the nonlinear map more complex. Conversely,
the use of 1D convolution could be effective when dealing with noisy datasets.

Other FF architectures named transformers [7] have recently shown outstanding
performances on NLP-related tasks, thanks to the so-called attention mechanism
[27]. Transformers are specifically designed for NLP (e.g., text translation from one
language to another, or question-answering), inherently different from the prediction of a numerical time series. These kinds of problems require an encoder which
compresses the information stored in the input sequence in a context vector, and a
decoder that, starting from the context vector, produces the output sequence. Encoder
and decoder have different parameters because their roles are inherently different. In
our neural predictors, on the other hand, inputs and outputs are samples of the same
variable, and thus the encoder-decoder architecture is unjustified.

Another issue is that the last value of the input sequence has to be processed twice:
it is the last input to the encoder, but also the first of the decoder. In principle, one
can adopt such structures in numerical time series forecasting [29], but it seems to
be a mind-bending solution. Transformers are always trained with TF and present a
feed-forward structure. In a certain sense, when considering a simple case such as

the logistic map (the single step prediction is equivalent to reproducing a parabola), transformers can this be seen as complex versions of the FF-recursive predictor.

6.5   Chaotic Dynamics in Recurrent Networks

Recurrent neural networks are dynamical systems which are known to frequently exhibit a chaotic behavior [14]. This is usually considered as a remarkable issue, especially in those applications (e.g., control) which require stability properties. For this reason, many authors proposed alternative formulations of the RNNs' equations and imposed apposite constraints to guarantee the model stability [18].

In the context considered in this work, we try to reproduce chaotic dynamics and thus imposing the stability of our neural model would be dangerous, since the system's behavior is not stable. We think that the RNNs' natural aptitude for being chaotic [16] would actually help them in reproducing the dynamics of the considered oscillators.

References

1. Bai, S., Zico Kolter, J., & Koltun, V. (2018). An empirical evaluation of generic convolutional and recurrent networks for sequence modeling. arXiv:1803.01271.
2. Bakker, R., et al. (2000). Learning chaotic attractors by neural networks. *Neural Computation, 12.10*, 2355–2383.
3. Bocquet, M., et al. (2020). Bayesian inference of chaotic dynamics by merging data assimilation, machine learning and expectation-maximization. *Foundations of Data Science, 2.1*, 55–80.
4. Brajard, J., et al. (2020). Combining data assimilation and machine learning to emulate a dynamical model from sparse and noisy observations: A case study with the Lorenz 96 model. *Journal of Computational Science, 44*, 101171.
5. Cuéllar, M. P., Delgado, M., & Pegalajar, M. C. (2007). An application of non-linear programming to train recurrent neural networks in time series prediction problems. In *Enterprise Information Systems VII* (pp. 95–102). Springer.
6. Dercole, F., Sangiorgio, M., & Schmirander, Y. (2020). An empirical assessment of the universality of ANNs to predict oscillatory time series. *IFAC-PapersOnLine, 53.2*, 1255–1260.
7. Devlin, J., et al. (2018). Bert: Pre-training of deep bidirectional transformers for language understanding. arXiv:1810.04805.
8. Farina, M., & Piroddi, L. (2010). An iterative algorithm for simulation error based identification of polynomial input-output models using multi-step prediction. *International Journal of Control, 83.7*, 1442–1456.
9. Farina, M., & Piroddi, L. (2012). Identification of polynomial input/output recursive models with simulation error minimisation methods. *International Journal of Systems Science, 43.2*, 319–333.
10. Farina, M., & Piroddi, L. (2011). Simulation error minimization identification based on multistage prediction. *International Journal of Adaptive Control and Signal Processing, 25.5*, 389–406.
11. Farina, M., & Piroddi, L. (2008). Some convergence properties of multi-step prediction error identification criteria. In *2008 47th IEEE Conference on Decision and Control* (pp. 756–761).

12. Galván, I.M., & Isasi, P. (2001). Multi-step learning rule for recurrent neural models: an application to time series forecasting. *Neural Processing Letters, 13.2*, 115–133.
13. Kumpati, S. N., Kannan, P. et al. (1990) Identification and control of dynamical systems using neural networks. *IEEE Transactions on neural networks, 1.1*, 4–27.
14. Laurent, T., & von Brecht, J. (2016). A recurrent neural network without chaos. arXiv:1612.06212.
15. LeCun, Y., Bengio, Y., et al. (1995). Convolutional networks for images, speech, and time series. In *The handbook of brain theory and neural networks* (Vol. 3361.10).
16. Li, Z., & Ravela, S. (2019). On neural learnability of chaotic dynamics. arXiv:1912.05081.
17. Menezes, J. M. P., Jr., & Barreto, G. A. (2008). Long-term time series prediction with the NARX network: An empirical evaluation. *Neurocomputing, 71.16-18*, 3335–3343.
18. Miller, J., & Hardt, M. (2018). Stable recurrent models. arXiv:1805.10369.
19. van den Oord, A., et al. (2016). Wavenet: A generative model for raw audio. arXiv:1609.03499.
20. Pancerasa, M., et al. (2018). Can advanced machine learning techniques help to reconstruct barn swallows' long-distance migratory paths? In *Artificial Intelligence International Conference*. PremC. pp. 89–89.
21. Pancerasa, M. et al. (2019). Reconstruction of long-distance bird migration routes using advanced machine learning techniques on geolocator data. *Journal of the Royal Society Interface 16.155*, 20190031.
22. Pathak, J., et al. (2017). Using machine learning to replicate chaotic attractors and calculate Lyapunov exponents from data. *Chaos: An Interdisciplinary Journal of Nonlinear Science, 27.12*, 121102.
23. Piroddi, L., & Spinelli, W. (2003). An identification algorithm for polynomial NARX models based on simulation error minimization. *International Journal of Control, 76.17*, 1767–1781.
24. Ribeiro, A. H., & Aguirre, L. A. (2018). Parallel training considered harmful?: Comparing series-parallel and parallel feedforward network training. *Neurocomputing, 316*, 222–231.
25. Sangiorgio, M. (2021). Deep learning in multi-step forecasting of chaotic dynamics.. Ph.D. thesis. Department of Electronics, Information and Bioengineering, Politecnico di Milano.
26. Sangiorgio, M., & Dercole, F. (2020) Robustness of LSTM neural networks for multi-step forecasting of chaotic time series. *Chaos, Solitons & Fractals, 139*, 110045.
27. Vaswani, A. et al. (2017). Attention is all you need. *Proceedings of the 31st Conference on Neural Information Processing Systems*, 30, 5998–6008.
28. Werbos, P. J. (1990). Backpropagation through time: What it does and how to do it. In *Proceedings of the IEEE* (Vol. 78.10, pp. 1550–1560).
29. Wu, N., et al. (2020). Deep Transformer Models for Time Series Forecasting: The Influenza Prevalence Case. arXiv:2001.08317.

Chapter 7
Concluding Remarks on Chaotic Dynamics' Forecasting

Abstract In this book, we compared different neural approaches in the forecasting of chaotic dynamics, which are well-known for their complex behaviors and the difficulty of their prediction. Our analysis shows that the LSTM predictor trained without teacher forcing is the most accurate approach in the forecasting of complex oscillatory time series. This predictor always provides the best accuracy in all the considered tasks, spanning a wide range of complexity and noise sources. It also demonstrates the ability to adapt to other domains with similar features without a relevant decrease of accuracy. The comparison with the real system used as predictor in a noisy environment is particularly interesting: even the complete knowledge of the system structure does not allow perfect predictions when the initial conditions are only approximately known. This allows the border between time series forecasting and system identification problems to be clearly defined.

In this book, we compared the forecasting accuracy and robustness of four different autoregressive predictors based on feed-forward and recurrent neural nets on artificial (both noise-free and noisy) time series generated by chaotic oscillators, and real-world datasets that exhibit a chaotic behavior. Specifically, we considered feed-forward (FF) recursive and multi-output nets, and a recurrent architecture with LSTM cells. Regarding the LSTM nets, we questioned the so-called teacher forcing (TF) training method in the context of time series forecasting. This method, successfully put forward in natural language processing applications, consists of training the network using the target from the prior time steps instead of feeding back the network predictions as the input for the following steps. Adopting TF, the network is not trained to compensate the effect of one-step prediction errors on multi-step predictions (exposure bias problem). In other words, the LSTM-TF focuses on the single-step forecasting, and consequently has the same issues of the FF-recursive predictor [9, 10]. We proposed a training method for RNNs that avoids TF (LSTM-no-TF in our figures) and we performed an in-depth comparison with TF and with the two FF predictors.

We showed that LSTM predictors outperform FF-recursive and multi-output ones in four standard chaotic systems: logistic, Hénon and two generalized Hénon maps with low and high dimension. The predictive power of LSTM predictors can be further enhanced by training without TF. The fact that the predictors' performances hold the same ranking despite the different complexity of the considered systems (the fractal dimension of their attractors spans from 0.5 to 9.13) proves the robustness of the obtained results. In addition, the results are in accordance with those obtained by recent works in the field on different case studies (see for instance [6–8, 13]). Machine learning tools almost perfectly predict the future evolution of the system for 5/6 Lyapunov times; after that, predictions start degrading significantly.

When analyzing the predictive power within the same chaotic attractor, the situation is quite similar to that described above: LSTM, especially when trained without TF, ensures a more uniform predictive power with respect to the two FF nets.

The presented results, shown for convenience in discrete-time maps, are expected to hold in a broader context, including continuous-time systems [2] and real-world datasets (e.g., signal related to meteorology, fluid dynamics, biometric, econometrics) [5, 12]. As we explained in Chap. 6, the extension to continuous-time systems is straightforward, as long as we handle the issues related to the numerical integration carefully. Conversely, extending the results to noisy and real-world datasets is conceptually much more challenging because of the exponential propagation of noise in a chaotic system.

To ensure a better interpretability of the results, we did not directly consider a real-world dataset. We first investigated the effect of observation noise by superimposing an additive artificial disturbance on the four deterministic time series mentioned above. The results obtained for different levels of noise confirmed the ranking already observed in the noise-free case: LSTM-no-TF is the highest performing predictor. This experiment also allowed us to prove that the model identification error is negligible if compared to the observation error even when the latter is quite limited (in the order of 0.5%). In other words, having the actual model of the system available is practically useless (at least in terms of predictive accuracy) if one can properly identify a neural predictor [11].

We then introduced a modified version of the logistic map by setting a slow periodic dynamic for the growth-rate parameter. This slow dynamic is somewhat similar to the annual cycle of environmental variables and can be seen as a structural disturbance, while the fast dynamic of the logistic plays the role of the rapid processes occurring in the atmosphere. In this case, the gap between LSTM and FF predictors is quite broad, suggesting that the dynamic nature of recurrent architectures is more suitable for modelling these time-varying processes than the static ones of the feedforward nets.

Finally, we considered solar irradiance and ozone concentration to evaluate how the situation changes when switching from artificial to real-world systems. Both time series derive from a chaotic system as proved by the presence of a positive Lyapunov exponent.

Generally speaking, the results reported in this study further confirm the well-known accuracy of FF and LSTM networks in predicting time series related to environmental variables [4]. LSTM-no-TF was confirmed to be the highest performing predictor. FF-multi-output provides a forecasting accuracy almost identical to LSTM-no-TF on a 12-hour horizon, while the gap between the two grows on the

48-step horizon. The FF-recursive and LSTM-TF predictors have a lower predictive power, especially on long-term forecasting horizons. Focusing on LSTM nets, the importance of the training approach used emerges also in real-world datasets.

Another interesting conclusion is that among the FF-recursive and FF-multi-output, the former is the one that exhibits the lowest performances when used in real-world datasets. As already reported, a rough explanation probably lies in the fact that its parameters are optimized over a time horizon of 1 step, but then used for a longer horizon, thus propagating the error. Therefore one of the merits of this study is to clarify that a common practice, namely to identify a single-step predictor and then use it in a recursive way to forecast the sequence $\hat{y}(t + 1), \hat{y}(t + 2), \ldots, \hat{y}(t + h)$, may not be the best choice. To this end, the proposed FF-multi-output may represent a more suitable alternative. Expanding this idea to a broader context, we can conclude that prediction and system identification are certainly related but different tasks. A FF network trained on the one-step prediction task can mimic the behavior of the real chaotic system fairly well, but other configurations exhibit greater predictive power and robustness over multiple steps.

The neural networks developed in this study, with particular focus on the LSTM-no-TF predictor, have proved able to forecast solar radiation in other stations of a relatively wide domain with a minimal loss of precision and without the need to retrain them. This opens the way for the development of county or regional predictors, valid for ungauged locations with different geographical settings. Such precision may perhaps be further improved by using other meteorological data as input to the model, thus extending the purely autoregressive approach adopted in this study. Ad-hoc training on each sequence would undoubtedly improve the performance, but the exact purpose of this analysis is to show the potential of networks calibrated on different stations and to evaluate the possibility of adopting a predictor developed elsewhere when a sufficiently long series of local values is missing. When lots of data are available for a certain location but that amount is still not enough to ensure a robust identification process of the neural predictor, an approach somewhat half-way between retraining from scratch and the domain adaptation strategy, named fine tuning in the machine learning literature, can be adopted. It consists of starting from the network parameters optimized for the source domain and retraining with a subset of the target domain data only the final layer of the net.

In all the experiments considered, it emerges that LSTM-no-TF provides the best performance. The distance between the latter and the other predictors is essentially task-dependent. The LSTM-no-TF network shows all the strengths of the three competitors: it reproduces the recursive evolution rule of the dynamical systems well (as FF-recursive), it is optimized over the entire forecasting horizon (as FF-multi-output), and its internal dynamics make it suitable for sequential tasks (as LSTM-TF).

An overview of the results obtained in terms of forecasting accuracy is shown in the radar chart reported in Fig. 7.1.

In addition, the LSTM-no-TF turned out to have the best robustness with respect to the network's hyperparameters, including the number of autoregressive terms given as the input (i.e., the dataset embedding dimension). Most of the time, the embedding dimension has to be estimated with a non-trivial process, as demonstrated by recent

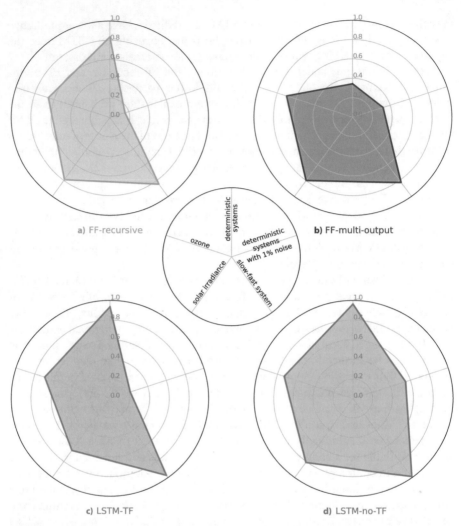

Fig. 7.1 Radar chart reporting the forecasting accuracy (R^2 score) obtained by each neural predictor in the five tasks. When a task is composed by multiple subcases (e.g., "deterministic systems" includes four chaotic systems), an average performance indicator is reported

work which tackled this issue (see, for instance, [1, 3]). The robustness with respect to this source of uncertainty is thus undoubtedly a remarkable advantage.

The experiment performed on the 3D generalized Hénon map (Fig. 5.5), for which we knew a-priori the correct number of input steps to be fed into the model, m, showed that LSTM predictors are more robust than their FF competitors. The same conclusion can be reached when analyzing the performances on the slow-fast logistic (note that in this case the actual value of m is unknown). The LSTM nets always outperform the two FF approaches.

These deep learning techniques have profoundly changed the notion of predictability as imagined by the first researchers and have opened new development lines that are rapidly expanding. The application of other network architectures, such as the Convolutional Neural Networks (CNNs), originally developed for image analyses, is just one example. We can also imagine future network structures that combine the recognition ability of CNNs with the dynamic evolution of LSTM. The evolution of parallel hardware will make the synthesis of very complex networks with thousands neurons faster and faster, and will possibly allow the development of automatic tools to help determining the most advisable structure. All these developments will push further our possibility of predicting complex oscillatory time series with sufficient accuracy. However, accurate long-term predictions will remain a chimera. As shown in the book, even if the perfect knowledge of the system was possible, this would not allow a long extension of our forecasting ability since an exact measure of the current condition of the system is not feasible.

References

1. Bradley, E., & Kantz, H. (2015). Nonlinear time-series analysis revisited. *Chaos: An Interdisciplinary Journal of Nonlinear Science, 25.9*, 097610.
2. Dercole, F., Sangiorgio, M., & Schmirander, Y. (2020). An empirical assessment of the universality of ANNs to predict oscillatory time series. *IFAC-PapersOnLine, 53.2*, 1255–1260.
3. Deshmukh, V., et al. (2020). Using curvature to select the time lag for delay reconstruction. *Chaos: An Interdisciplinary Journal of Nonlinear Science, 30.6*, 063143.
4. Guariso, G., Nunnari, G., & Sangiorgio, M. (2020). Multi-Step Solar Irradiance forecasting and domain adaptation of deep neural networks. *Energies, 13.15*, 3987.
5. Matsumoto, T. et al. (2001). Reconstructions and predictions of nonlinear dynamical systems: A hierarchical Bayesian approach. *IEEE Transactions on Signal Processing, 49.9*, 2138–2155.
6. Pathak, J., et al. (2018). Hybrid forecasting of chaotic processes: Using machine learning in conjunction with a knowledge-based model. *Chaos: An Interdisciplinary Journal of Nonlinear Science, 28.4*, 041101.
7. Pathak, J., et al. (2018). Model-free prediction of large spatiotemporally chaotic systems from data: A reservoir computing approach. *Physical Review Letters, 120.2*, 024102.
8. Pathak, J., et al. (2017). Using machine learning to replicate chaotic attractors and calculate Lyapunov exponents from data. *Chaos: An Interdisciplinary Journal of Nonlinear Science, 27.12*, 121102.
9. Sangiorgio, M. (2021) Deep learning in multi-step forecasting of chaotic dynamics. Ph.D. thesis. Department of Electronics, Information and Bioengineering, Politecnico di Milano.
10. Sangiorgio, M., & Dercole. (2020). Robustness of LSTM neural networks for multi-step forecasting of chaotic time series. *Chaos, Solitons and Fractals, 139*, 110045.
11. Sangiorgio, M., Dercole, F., & Guariso, G. (2021). Forecasting of noisy chaotic systems with deep neural networks. *Chaos, Solitons & Fractals, 153*, 111570.
12. Siek, M. (2011). Predicting Storm Surges: Chaos, Computational Intelligence, Data Assimilation and Ensembles: UNESCO-IHE Ph.D. thesis. CRC Press.
13. Vlachas, P. R., et al. (2020). Backpropagation algorithms and Reservoir Computing in Recurrent Neural Networks for the forecasting of complex spatiotemporal dynamics. *Neural Networks, 126*, 191–217.

Index

A
Attractor, 3, 12, 98
   chaotic, 14, 18, 19, 64, 87, 98
   cycle, 13
   fractals, 23
   quasi-periodic, 13, 26
   strange, 12, 13, 17, 19, 23
   torus, 13
   weakly chaotic, 18, 20
Attractor reconstruction, 23
Autonomous system, 12, 43

B
Backpropagation, 91, 94
   through time, 91, 94
Basin of attraction, 13, 17, 22
Boundedness, 11, 19, 22
Butterfly effect, 11

C
Clear sky index, 74, 75
Correlation dimension, 21, 73

D
Delay-coordinate embedding, 24, 64, 66
Delay coordinates, 24
Dense trajectory, 13, 20
Diffeomorphism, 24
Domain adaptation, 4, 77, 99
Dynamical system, 3, 12, 23, 31, 54, 69, 87, 95, 99

E
Eigenvalue, 15
Eigenvector, 15, 18
Embedding dimension, 4, 24, 27, 43, 44, 65, 66, 73, 78, 86, 99
Exposure bias, 3, 48, 62, 97

F
False nearest neighbor algorithm, 25, 66, 73
Fine tuning, 99
Fixed point, 13
Folding, 12, 19
Fractal dimension, 21, 34, 44, 66, 87, 98
Fractal structure, 20, 23
Fundamental solution matrix, 14

H
Hénon map, 33, 59, 90, 97
   generalized, 33, 59, 60, 63, 64, 90, 97
Hyperchaos, 19, 31, 33
Hyperparameter, 55, 56, 65, 99

I
Identification error, 69, 98

J
Jacobian matrix, 14, 18, 27

K
Kaplan-Yorke formula, 22–24, 34, 66
Keras, 45, 52

L
Least squares, 94
Logistic map, 23, 32, 35, 59, 64, 90, 95, 97
Long-term climate, 55, 85, 87
Lorenz, Edward, 11
Lyapunov dimension, 22
Lyapunov exponents, 12, 14, 16, 34, 55
    largest, 17, 19, 27, 53, 64, 73, 87
    local, 18
    non-leading, 17
    partial sums, 17
Lyapunov time, 21, 28, 53, 60, 64, 69, 85, 98

M
Mean squared error, 52, 54
Multi-step forecasting, 87, 90

N
Noise-free, 26, 59, 67, 94, 97, 98
Noisy, 26, 59, 67, 73, 94, 97, 98
Nonlinear regression, 34, 44
Non-stationary, 4, 34, 71
Numerical integration, 85, 98

O
Observation error, 69, 98
Observation noise, 4, 67, 98
Ozone concentration, 4, 37, 55, 59, 73, 78,
    98

P
Parameters sharing, 47, 86
Persistent model, 74, 75, 80, 81
PyTorch, 45, 48, 52, 86

Q
Quasi-periodic trajectory, 20, 21

R
Repeller, 13
Reservoir computing, 2, 94
Reversible system, 12, 15, 18
Robustness, 100

S
Saddle, 13
Sensitivity to initial conditions, 11, 14, 19
Singular value, 15, 18
Slow-fast system, 35, 71, 98
Solar irradiance, 4, 36, 55, 59, 73, 74, 93, 98,
    99
Spectrum, 20
Stable manifold, 13, 17
Stochastic, 67
Stretching, 12, 19
Structural noise, 4, 71
Supervised learning, 4, 43
System identification, 87, 90, 93, 99

T
Takens' theorem, 24, 26, 65
Teacher forcing, 3, 47, 59, 75, 91, 93, 97
TensorFlow, 45, 48, 52
Time-independent system, 12
Transfer learning, 4
Transformers, 94

U
Unfolding, 52, 91

V
Variational equation, 14

Printed in the United States
by Baker & Taylor Publisher Services